高等院校土建学科双语教材（中英文对照）
◆ 建筑学专业 ◆
BASICS

计算机辅助设计
CAD

[德] 扬·凯博斯　编著
杨　雷　刘婷婷　译

U0269846

中国建筑工业出版社

著作权合同登记图字：01-2007-3330 号

图书在版编目（CIP）数据

计算机辅助设计/（德）凯博斯编著；杨雷，刘婷婷译.—北京：中国建筑工业出版社，2011
高等院校土建学科双语教材（中英文对照）◆建筑学专业◆
ISBN 978-7-112-11585-3

Ⅰ.计… Ⅱ.①凯…②杨…③刘… Ⅲ.计算机辅助设计-高等学校-教材-汉、英 Ⅳ.TP391.72

中国版本图书馆 CIP 数据核字（2009）第 209817 号

Basics：CAD/Jan Krebs
Copyright © 2007 Birkhäuser Verlag AG（Verlag für Architektur），P. O. Box 133，4010 Basel，Switzerland
Chinese Translation Copyright © 2011 China Architecture & Building Press
All rights reserved.
本书经 Birkhäuser Verlag AG 出版社授权我社翻译出版

责任编辑：孙　炼
责任设计：郑秋菊
责任校对：王金珠　兰曼利

高等院校土建学科双语教材（中英文对照）
◆建筑学专业◆
计算机辅助设计
［德］扬·凯博斯　编著
杨　雷　刘婷婷　译
*
中国建筑工业出版社出版、发行（北京西郊百万庄）
各地新华书店、建筑书店经销
北京嘉泰利德公司制版
北京建筑工业印刷厂印刷
*

开本：880×1230 毫米　1/32　印张：5⅛　字数：165 千字
2011 年 5 月第一版　2011 年 5 月第一次印刷
定价：**18.00** 元
ISBN 978-7-112-11585-3
　　　（20272）

版权所有　翻印必究
如有印装质量问题，可寄本社退换
（邮政编码 100037）

中文部分目录

\\ 序   6

\\ CAD：定义及应用领域   93

\\ 虚拟绘图板   94
　　\\ 用户界面   94
　　\\ 坐标系   97
　　\\ 透明的平面——分层原则   99

\\ 绘图功能   104
　　\\ 绘图元素   104
　　\\ 设计工具   110
　　\\ 修改功能   114

\\ 三维绘图   119
　　\\ 三维设计   119
　　\\ 构图方法   124
　　\\ 建模   128
　　\\ 建筑元素   130

\\ 视觉效果   137
　　\\ 表面   141
　　\\ 光影处理   143
　　\\ 透视图和虚拟相机   145
　　\\ 渲染参数   148

\\ 应用数据集   151
　　\\ 程式库   151
　　\\ CAD 对外接口   152
　　\\ TAI   153
　　\\ 打印和制图   155

\\ 系统需求　157
　　\\ 硬件设备　157
　　\\ 软件　158

\\ 与电脑交互进行设计　159

\\ 附录　160
　　\\ 软件要点表　160
　　\\ 图片来源　163

CONTENTS

\\Foreword _7

\\CAD: Meaning and fields of application _8

\\The virtual drawing board _10
    \\User interface _11
    \\Coordinate systems _15
    \\Transparent planes – the layering principle _17

\\Drawing functions _22
    \\Drawing elements _23
    \\Design tools _30
    \\Modifications _34

\\The third dimension _40
    \\Three-dimensional design _40
    \\Construction methods _48
    \\Modelling _52
    \\Architectural elements _54

\\Visualization _62
    \\Surfaces _66
    \\Light and shade _69
    \\Perspective and virtual camera _71
    \\Rendering parameters _75

\\Data flow _78
    \\Program libraries _78
    \\CAD interfaces _80
    \\TAI _81
    \\Printing and plotting _82

\\System requirements _85
    \\Hardware _85
    \\Software _86

\\Designing in dialogue with the computer _88

\\Appendix _89
    \\Checklists and overview of software _89
    \\Picture credits _92

序

如果没有电脑的帮助，我们无法想像今天的设计制图工作将会是什么样子。在复杂的工程项目中，各个专业之间顺畅的数据交换和设计修改统筹都在电脑的介入帮助下得以最终实现。在过去的几十年里，计算机辅助设计（CAD）已经成为设计师的必备工具。不管结构复杂的设计还是简单的设计，CAD工具都可以帮助设计师创建出完整全面的设计资料。CAD也可以模拟显示状态以及提供详尽的参数信息。目前，几乎所有的建筑及规划单位都在应用CAD工具，各个大学里也有很多此类工具用于教学及研究。

目前市面上可以买到很多关于CAD工具功能及使用的手册，而这些书籍对CAD工具的基本原理却很少详细说明，而恰恰是这些基于基本原理且看似简单的说明和小窍门更能够帮助初学者免于养成很多日后难以纠正的不良习惯——一本书极具趣味性地向初学者讲解了CAD工具的功能、流程及构成等方面。本书的主要目标读者群为建筑学、工程、园林设计以及室内装潢等专业大学一年级的学生，同样，刚刚进入建筑制图或工程制图领域且今后必将以CAD为主要生产工具的人们亦在其列。

编者：贝尔特·比勒费尔德（Bert Bielefeld）

FOREWORD

Creating design and construction drawings would be inconceivable today without the help of computers. The complexity of major building projects has made it necessary for all participants to be able to exchange data smoothly and to modify layouts regularly in the process of creating the drawing that will ultimately be implemented. Over the past few decades, computer-aided design (CAD) has evolved into a universal tool for planners. CAD programs make it possible to create all documentation, from simple designs to complex construction drawings. They can also display photorealistic depictions or information on quantities and components. Nowadays nearly all architectural and planning offices use the full range of CAD features, and it has become a firm fixture in university studies as well.

Whereas it is possible to buy handbooks and publications that describe the functions of specific CAD programs, what is lacking is a general survey of the basic principles of CAD. Many tips and tricks would be particularly helpful for CAD beginners, keeping them from making basic mistakes that require time-consuming corrections. This is the gap that the Basics student series intends to fill with *Basics CAD*, which provides a fundamental understanding of the functions, processes and structures that all programs share. *Basics CAD* targets first-year students in disciplines such as architecture, engineering, landscape design and interior architecture, as well as trainees in the fields of construction drawing and technical drawing who will work with CAD programs in their later professional environments.

Bert Bielefeld, Editor

CAD: MEANING AND FIELDS OF APPLICATION

CAD – short for computer-aided design – is the catchword for drawing and designing with the help of a computer. A wide selection of both simple and complex CAD computer programs (software) makes it possible to create two- and three-dimensional drawings using input devices such as the keyboard, mouse and other tools. These drawings are displayed by output devices such as monitors and printers. The CAD field of IT plays an important role in many areas including engineering. It is used in automotive and plant engineering, as well as in structural engineering and architecture – the focus of this book. Aside from creating technical drawings, CAD programs can be used to develop powerful virtual models that provide a basis for a wide range of simulations. CAD programs can generate photorealistic visualizations of buildings, as well as climate and light simulations. With the right CAD software, users can also design load and fluid flow simulations in addition to simplifying development and production processes.

The first CAD programs were developed in the 1960s and were used primarily in aeroplane construction. With the breakthrough of personal computing and the lower costs of computer workstations in the 1980s, these programs became available to a larger number of users. The powerful, standardized computing systems developed in the early twenty-first century has led to efficient, relatively inexpensive CAD systems that meet a wide range of requirements.

Although nearly all the programs have a similar foundation, they differ significantly at times in terms of operation and use. There is a great deal of literature on CAD on the market – so much that it is nearly impos-

> \\Hint:
> In order to establish a practical relationship to application software, Basics CAD includes a selection of sample functions from leading CAD software providers in the CAAD (computer-aided architectural design) field. You can find an extensive list of the diverse CAD programs on offer in the appendix of this book.

sible to keep track of – but most of it deals with particular CAD systems. By contrast, the book *Basics CAD* offers a general approach to learning CAD. It contains basic, application-oriented knowledge and is designed to assist the novice in selecting a suitable CAD system.

THE VIRTUAL DRAWING BOARD

When you draw with CAD, the sheet of paper is replaced by the computer monitor, the pencil by the mouse and keyboard. Lines and shapes are created by mouse clicks and keyboard entries, and the whole process is supported by functions that simplify drawing. > Chapter Drawing elements

Screen scale

The objects in CAD drawings are usually created on a scale of 1:1 – that is, at their real size. This means that a 10 m wall is drawn with a length of 10 m. To display the entire length of the wall on the screen, you must choose a smaller scale since the space would otherwise not be sufficient to show it. <u>Screen scale</u> describes the ratio between the real size of drawn objects and their depiction on the computer monitor. It changes automatically when the size of the screenshot changes. The actual representational scale that you follow when drawing on paper is created only in the output process in CAD (e.g. printing), but you must always bear it in mind when doing your drawing.

Reference scale

In a CAD drawing on the monitor, this representational scale is referred to as the reference scale. It describes the scale in which the drawn object is likely to be printed (e.g. 1:20, 1:100 or 1:500). Since some of the elements in a drawing, such as lettering, are independent of building components and must be displayed in an appropriate size in subsequent prints, they are not based on a drawing scale of 1:1, but on the probable scale of the printout. This means that, in the drawing operation, the software illustrates on screen the size ratio between the typeface and other elements, including the actual drawn object. > Chapter Printing and plotting

\\Tip:
Screen scale must be the same as the reference scale if you wish to view and evaluate the real size of the printout on the screen. Some CAD programs show the screen scale as a percentage of the reference scale: when set at 100% it thus shows the real size of the print. It is also advisable to create sample prints during the work process in order to evaluate the effects of the scale drawing.

\\Hint:
In addition to the mouse and keyboard, you can use a variety of other input devices, including sketchpad and SpaceMouse. However, these play a minor role in architecture-related CAD (see Chapter Hardware).

Measurement units

When technical drawings are made, architects use common units of length so that their work is comprehensible to everyone. In Central Europe, these units are millimetres, centimetres and metres. Some English-speaking countries continue to use feet and inches as units of length, but due to the need for standardization in managing international projects, even these countries are increasingly using the more easily convertible units favoured in Central Europe. You are free to select the units on which you wish to base your CAD drawing. The smallest units can be millimetres, centimetres or metres, or inches or feet.

USER INTERFACE

CAD user interfaces consist of special building blocks that we will explain briefly below by elucidating their underlying concepts.

The various tools of the CAD program are displayed by symbols and menus on the user interface. › Fig. 1 You can generate a drawing on the computer using the usual entry devices such as the mouse and the keyboard.

Drawing area

The drawing area is the most important part of the user interface. It allows you to draw and modify objects in either a two-dimensional field or in a space defined in three dimensions. Generally speaking, the CAD drawing area is comparable to a piece of paper. The main difference is that it is a virtual workspace that offers far-ranging options and various virtual tools.

Selection and drawing tools

The computer mouse is a commonly used <u>pointing device</u>. A pointing device controls the cursor on the user interface and functions as a virtual selection and drawing tool. It is represented by an arrow, crosshairs or some other symbol, according to the CAD program in question. This symbol generally changes when you use different selection and drawing options in the drawing operation to indicate the function you are currently using.

Electronic pens

CAD programs allow you to customize electronic pens and specify various line widths, line types and colours. › Fig. 2 You can define line properties before you begin to draw a line, or you can modify them retroactively. This brings greater clarity to both the drawing process and the methodology, since different lines emphasize and illustrate different elements of a drawing. Since many CAD programs link line colours to specific line widths, you can immediately see which elements have been drawn with a specific line width even in complex drawings. In this way, the different elements of a drawing are prepared for subsequent printing based on their

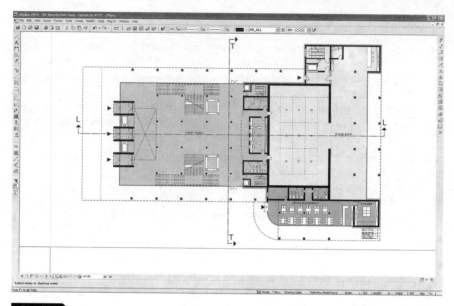

Fig.1:
Example of a CAD interface (Nemetschek Allplan)

screen depiction. › Chapter Printing and plotting The depiction of the pen line represents the subsequent result in the printout.

In addition, many CAD programs allow you to show true-to-scale line widths on the screen. In the corresponding screen scale, the line widths are shown just as they appear in a printout and give an impression of what the print will actually look like. Electronic pens can also be assigned other virtual properties that are important for the visualization of structures.
› Chapter Visualization

Menu bar

The menu bar is located above the drawing area and displays the software options in interactive lists. For instance, as is generally the case with Windows-based applications, you can use the "File" menu to create new drawings, open existing ones or store drawing files. Via the menu bar, you can also custom-configure the user interface and activate all software commands and functions by means of the mouse and the keyboard.

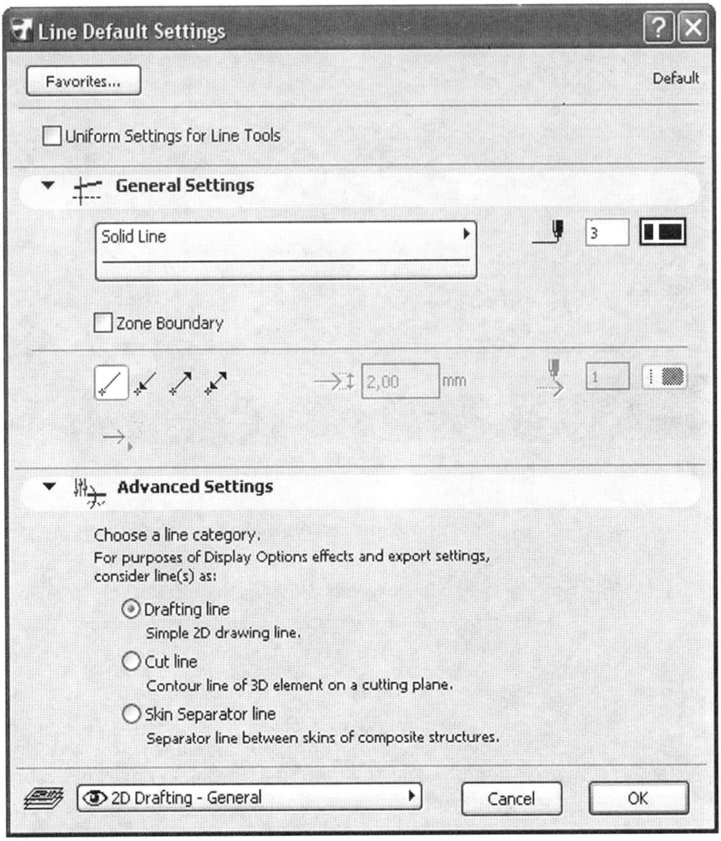

Fig. 2:
You can customize the width, type and colour of the lines drawn by electronic pens (Graphisoft ArchiCAD).

Toolbox

The toolbox buttons display symbols of the drawing and tool functions available. They are activated by selection with the virtual selection tool. › Fig. 3 If you move the selection tool to a symbol, the software will usually display the name of the corresponding function. When you use a CAD program for the first time, the user interface will contain default functions that can be configured individually and shown or hidden according to your needs.

13

Fig.3:
Example of a toolbox with various functions (Graphisoft ArchiCAD)

Context menu   Context menus are related to the function you are currently using and contain commands relevant to each drawing operation. In default mode, they can be displayed by right-clicking the mouse, which will make them appear near the drawing and selection tool. Context menus allow you to repeat or cancel commands quickly, and to activate other drawing tools.

Dialogue boxes   A dialogue box allows you to engage in a "dialogue" with the CAD program on a specific function. Dialogue boxes often provide a more detailed description of a selected command, or explain the steps necessary to perform this command. They also make it possible to select special options

\\Tip:
The toolboxes can be displayed in a vertical or horizontal column on the edge of the user interface, or they can be positioned alongside or within the drawing area. An intelligent setup will enlarge the working area and allow you to work more efficiently. Once you have gained some experience of drawing with CAD, you will learn which functions you use frequently, and can then arrange the required toolboxes on the interface so that you can access them quickly.

\\Hint:
Cartesian coordinates are not the only available system. One alternative is the polar coordinate system, in which a point is described by a radius and an angle instead of by X and Y axes.

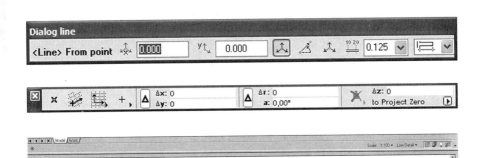

Fig. 4:
Communicating with CAD software through dialogue boxes (Nemetschek Allplan, Graphisoft ArchiCAD, Autodesk Architectural Desktop)

or to enter numerical values via the keyboard. The boxes are part of the various drawing and tool functions, and they open automatically on the user interface when you activate the commands. › Fig. 4

COORDINATE SYSTEMS

The basic geometric reference system in CAD software is a coordinate system that defines the virtual drawing area as a construction plane. This construction plane can be thought of as a piece of graph paper divided by horizontal and vertical lines. Points on a plane, which are specified as coordinates, are thus clearly mapped out within the CAD system and can be used to define the location and shape of drawing elements. › Chapter Drawing elements

The Cartesian coordinate system, which is used most commonly, is based on two perpendicular axes (X and Y) in two-dimensional space. These axes describe the distance between any given point and the zero point of the system. › Fig. 5

Imaginary lines parallel to the X and Y axes intersect within the coordinate system and define the position of any given point.

Going one step further, we can define a line in geometric terms as the connection between two points. This is also how both endpoints of a line are specified in the Cartesian coordinate system. › Fig. 6

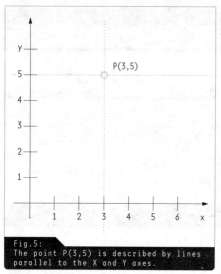
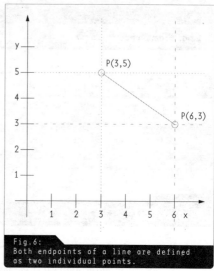

Fig.5:
The point P(3,5) is described by lines parallel to the X and Y axes.

Fig.6:
Both endpoints of a line are defined as two individual points.

Absolute coordinates

Absolute coordinates are based on the above-mentioned zero point of the drawing, at which the X and Y axes intersect. It makes sense to use absolute coordinates when the exact X and Y values of the coordinates you wish to define are known.

User-defined coordinate system

It would be disadvantageous to enter only absolute coordinates together with an absolute, fixed zero point since this would make large drawings particularly difficult to generate using the keyboard.

This is one of the main reasons that most CAD programs allow you to move and thus redefine the point of origin within the coordinate system

\\Hint:
A coordinate system like the one introduced in Figs 5 and 6 is not usually visible on the screen and serves only as an imagined or virtual basis for drawing. In some CAD programs, symbols appear in one corner of the drawing area for orientation purposes and to show the directions of the coordinate axes. They are, however, an exception.

\\Example:
Apartment walls can be drawn on one layer of a floor plan, the interior furnishings on another. Both levels can be viewed separately, edited or, if necessary, combined (see Fig. 8).

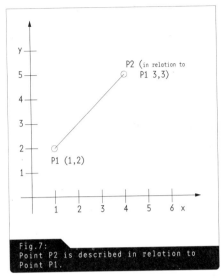

Fig.7:
Point P2 is described in relation to Point P1.

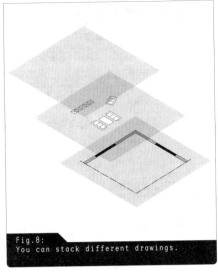

Fig.8:
You can stack different drawings.

any time you draw new elements. You may also rotate the coordinate system and thus the entire construction plane by any angle of your choosing. This makes it easy to design linear objects that are not perpendicular to the original coordinate system. Below, we refer to such user-defined coordinate systems as UCS.

Relative coordinates

A flexible, alternative method for specifying coordinates incorporates so-called relative coordinates. These are based on the point last defined in the drawing process, which then becomes the temporary reference point for new coordinates. › Fig. 7

TRANSPARENT PLANES – THE LAYERING PRINCIPLE

CAD drawings often contain large amounts of information. As the scope of a drawing expands, so does its complexity, and you might find it increasingly difficult to work on or change elements. This is particularly problematic when many different objects are positioned right next to or on top of one another. CAD programs solve this problem by using variable drawing layers that, like transparent drawing paper, are placed on top of one another as needed. › Fig. 8 These layers may have different names in the various CAD programs. In Nemetschek Allplan, for example, they are called DRAWING FILES.

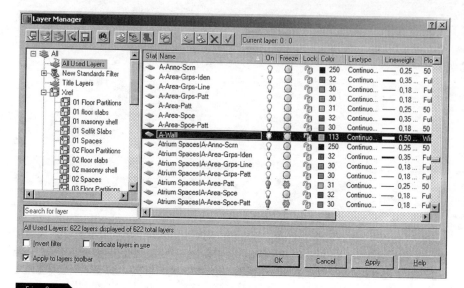

Fig.9:
Example of a CAD layer structure, which enables users to manage different layers and pen properties in a dialogue box (Autodesk Architectural Desktop).

\\Tip:
At the outset of a project, it is worthwhile to give some thought to the way layers — and thus the entire project — need to be organized. For instance, layers can be predefined for different drawing content, even if the layers are used only at a later stage in the project. In addition, if a component consists of several layers, they can be systematically combined to form meaningful overarching layer groups. This approach will allow you to create a hierarchy within a drawing, one that brings the requisite clarity even to complex projects (see Fig. 10).

\\Example:
If two layers are stacked on top of each other, it is possible to see both, but it makes sense to lock one while you continue to work on the other so that you do not inadvertently alter the first one. This is particularly important for complex, multi-layered drawings.

\\Hint:
With many CAD programs (e.g. Graphisoft ArchiCAD), you can structure a drawing not only by layers but also by floors, a feature that is especially useful when designing buildings. The floors are linked to specific elevations and permit a vertical organization of floor plans (based on ground floor, first floor, second floor etc.).

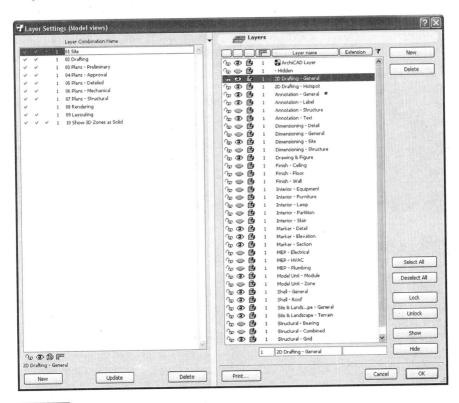

Fig.10:
Creating a hierarchy for drawing content within the layer structure. This brings clarity to the project (Graphisoft ArchiCAD).

In addition, layer structure plays an important role in organizing a drawing. Using a special dialogue box, you can create any number of layers and manage different drawing content easily. If a structure has many different design elements, this considerably simplifies the design process: independent layers can be created for exterior and interior walls, dimension lines, text, hatching and so on. You can organize drawing contents and retrieve drawings easily. > Fig. 9 and Appendix, Table 1

Further, by creating specific layers, you can show or hide drawing content, or you can "lock" a drawing, making it impossible to alter for a defined period.

Filter function   The layer system described above not only creates clarity in drawing operations but also allows you to retrieve specific drawing components or general elements quickly and easily when needed. That said, all categorization systems have logical limits or, in certain situations, are not consistently followed. For this reason, many CAD programs also offer a special function that permits users to search for special properties of drawing elements, by either filtering these elements out or activating them. › Fig. 11 This filter function goes by different names in current CAD systems and is sometimes referred to as the FILTER ASSISTANT, or by the more pragmatic term SEARCH AND ACTIVATE. It is normally selected from the menu bar.

\\ Example:
If you have drawn all the exterior walls in a complex floor plan using a special line type, which then proves to be too thick or too thin on a sample print, you can filter out the walls based on their line properties and redefine all of them in just a few short steps. They do not have to be selected individually.

\\ Tip:
The filter function can be used for a great variety of CAD elements. Search criteria include line colour and type, and component specifications (see Chapter Architectural elements). It is useful to draw certain elements or entire components using a specific line colour so that they can easily be filtered out and edited later on.

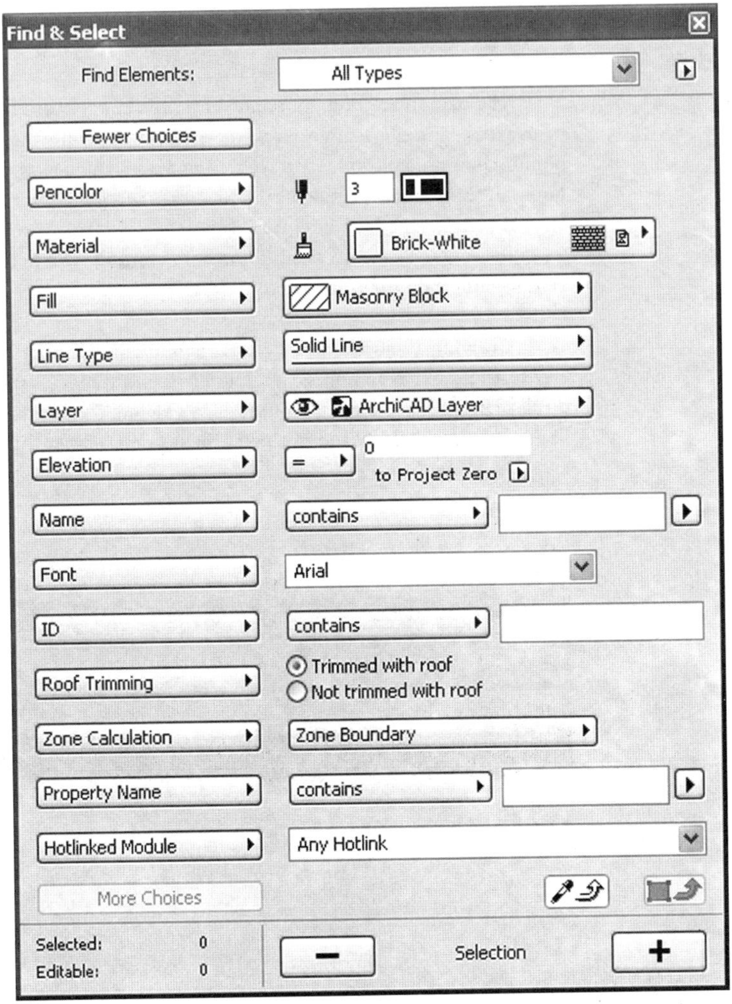

Fig.11:
Various filter criteria can be used to search for and activate drawing elements (Graphisoft ArchiCAD).

DRAWING FUNCTIONS

The design methods introduced in this chapter are basic approaches to drawing sample forms using the related drawing functions. They are popular, simple methods, and we will not treat them exhaustively or touch upon all their variations. Other approaches are possible, depending on the specific features of the program. In CAD there are almost always several ways to reach a destination, and it is important for you to find the quickest, simplest method for your own drawing.

Drawing is supported by CAD commands that make it possible to create the most common geometric elements (points, lines, squares and circles) in an easy, direct process. The various functions can be selected via the symbols displayed in the toolboxes on the user interface. › Fig. 12 and Chapter User interface

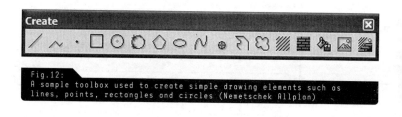

Fig.12:
A sample toolbox used to create simple drawing elements such as lines, points, rectangles and circles (Nemetschek Allplan)

\\Tip:
In addition to toolboxes and interactive menus, it is more efficient to use shortcuts to activate drawing and tool functions. Shortcuts are special keyboard combinations linked to specific functions. A well-known shortcut in the Windows operating system is CTRL+C, which copies a file onto the clipboard. The various CAD systems feature a number of default shortcuts that can be quickly adapted to individual needs via the menu bar.

\\Hint:
A polygon is a many-sided object that consists of an enclosed area and a specific number of edges. A polyline is a series of lines and vertices.

DRAWING ELEMENTS

Points

As in a hand drawing, the most basic element in a CAD drawing is a point. In geometric terms, a point is a zero-dimensional object that does not extend into space. All other geometric objects can be described by points and incorporate a particular number of them. For example, two points define a line, three points a triangle, four points a rectangle, eight points a cube and so on. Points are depicted through different symbols in the various CAD programs. › Fig. 13

To draw a point, you must first activate the corresponding drawing function using the POINT command.

With a simple click of the mouse, you can define the position of a point on the construction plane, or you can use the keyboard to enter the X and Y coordinates into the dialogue box. These drawing and entry principles provide a foundation for all the drawing functions described below, and they make it possible to create any desired shape – though at times a number of intermediate steps may be involved.

Lines

A simple straight line is described by two points. After you fix a point on the construction plane using the LINE command, you need to specify a second point as the endpoint. The CAD program then creates the line in the desired position. › Fig. 14

Polylines

A polyline consists of several line segments connected to create a coherent object. › Fig. 15

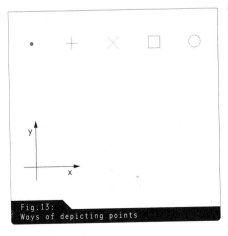

Fig. 13:
Ways of depicting points

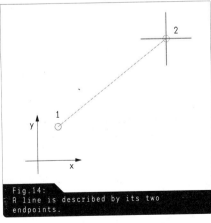

Fig. 14:
A line is described by its two endpoints.

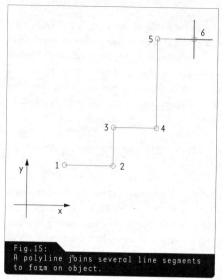

Fig.15:
A polyline joins several line segments to form an object.

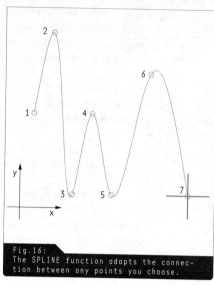

Fig.16:
The SPLINE function adapts the connection between any points you choose.

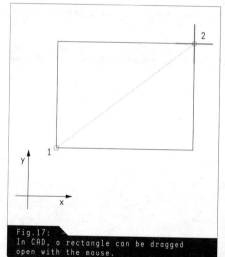

Fig.17:
In CAD, a rectangle can be dragged open with the mouse.

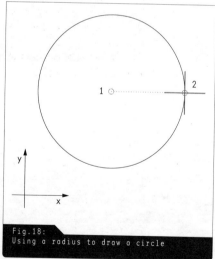

Fig.18:
Using a radius to draw a circle

Splines

The term "spline" comes from shipbuilding, where it is used to describe a pliant board that is anchored at several points and bent to fit the curvature of a ship's hull. The corresponding drawing function works in

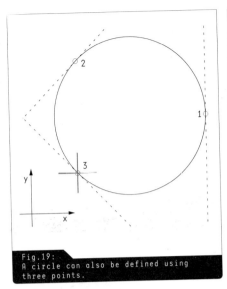 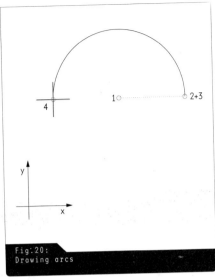

Fig. 19:
A circle can also be defined using three points.

Fig. 20:
Drawing arcs

the same way. It interpolates curves between arbitrary points, creating soft, flowing shapes. › Fig. 16

Squares and rectangles

A rectangle does not need to be drawn in a complicated fashion using four individual lines. It can be created much more easily via a single function. One method is to draw the rectangle using its diagonal. › Fig. 17 You can either drag it open with the mouse or enter its parameters with the keyboard.

Circles

There are various ways to define a circle. To begin with, you can describe it by its radius. In this case, you enter its centre into the dialogue box of the CIRCLE command. Then you specify the desired radius with the mouse-controlled drawing tool or enter a numerical value into the dialogue line using the keyboard. › Fig. 18 In other CAD programs, you can draw a circle by defining its diameter or three points on its perimeter, by using tangents, or by combining these methods of specifying points on a plane. › Fig. 19

Circular arcs

The arc is another important drawing element. Using the CIRCLE command, you can create either an entire circle or a section of it. If you wish to draw an arc, first define the centre and radius, then specify the arc's beginning and endpoint. › Fig. 20

25

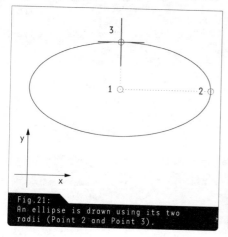

Fig. 21:
An ellipse is drawn using its two radii (Point 2 and Point 3).

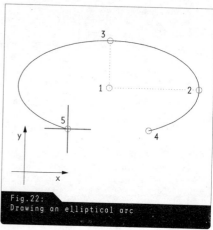

Fig. 22:
Drawing an elliptical arc

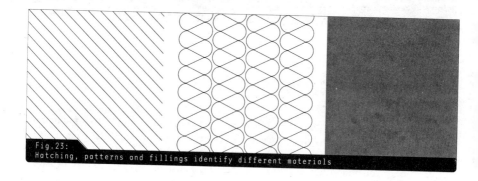

Fig. 23:
Hatching, patterns and fillings identify different materials

Ellipses

Ellipses are drawn in the same way as circles. The only difference is that both radii must be specified after the centre is defined. > Fig. 21

- Elliptical arcs

An elliptical arc is created in the same way as an ellipse, but requires a beginning and an endpoint as well. > Fig. 22

Hatching, patterns and fillings

Using hatching, patterns and fillings makes it easy to represent surfaces and enhances the legibility of individual drawing elements. > Fig. 23 Hatching provides information on the properties of both the materials and building components used in the design. Patterns and fillings allow an abstract depiction of surfaces and can also be used for the graphical pres-

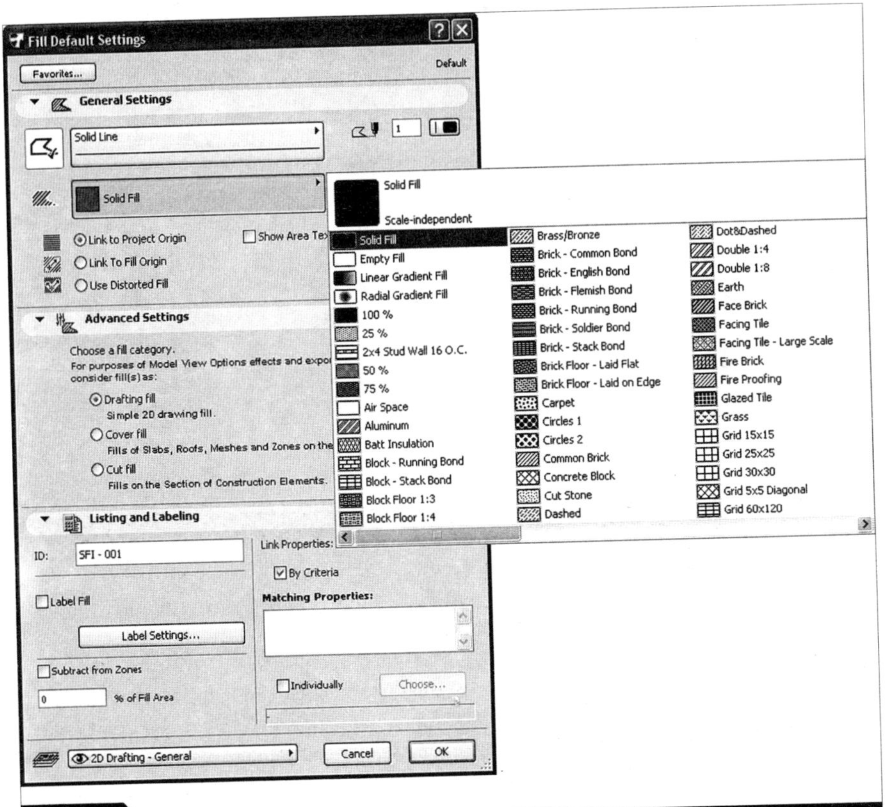

Fig.24:
A toolbox for creating hatches, patterns and fillings (Graphisoft ArchiCAD)

entation of drawings. Not all CAD programs distinguish between hatching, patterns and fillings. Many display all these functions in a single toolbox.
> Fig. 24

After selecting the desired surface by clicking the matching symbol, it can be "drawn" in a number of ways. You can enter each corner point of the depicted surface one after the other, or generate a rectangular surface using its diagonal, as in the process of creating a square. > page 25

Automatic boundary detection

Many CAD programs offer an alternative method for enclosed figures: if the outer perimeter of an object completely surrounds its inner

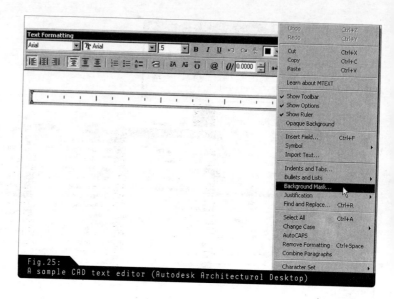

Fig. 25:
A sample CAD text editor (Autodesk Architectural Desktop)

 area, you can activate the <u>automatic boundary detection</u> function, which is usually located on the menu bar.

After this function is activated, the software inspects the surrounding contour line and fills in the selected surface. This saves a great deal of time, particularly when objects have complex geometries or a large number of corners.

Text elements

In many CAD programs, you can add text to drawings using a text editor. Like the common word-processing programs, this editor can be

\\Tip:
Automatic boundary detection works only for completely enclosed areas. If a drawing is imprecise, the contour lines may have small gaps that might not be visible when displayed on the screen in a normal size. Subsequent attempts to find the error can be time-consuming. For this reason – and for other drawing functions that build on one another – it is important to create precise drawings if you want to efficiently exploit the benefits of CAD.

\\Hint:
A precisely defined reference elevation, on which all higher or lower elevation points are based, is essential to ensure the correct relationship between the elevation points. This reference elevation is normally represented by the finished floor level (FFL) of the building's ground floor, and it equals ±0.00 m.
Additional information on dimension lines in particular and technical drawing in general can be found in Basics Technical Drawing by Bert Bielefeld and Isabella Skiba, Birkhäuser Publishers, Basel, 2006.

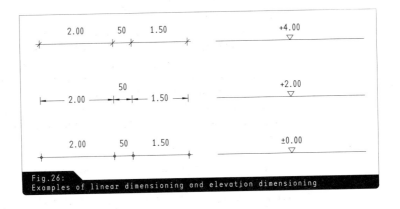

Fig. 26:
Examples of linear dimensioning and elevation dimensioning

used to change various parameters, including font type and character size. › Fig. 25 After activating the editor, you enter text word by word or copy it from other programs (e.g. Microsoft Word) using the standard copy function. However, some CAD programs do not have a text editor with a special entry window. In this case, you enter text directly into the drawing area and select it there for subsequent editing.

Dimension lines

Even though technical drawings are usually drawn to a particular scale, you should define all the relevant dimensions of a drawing. This can be done with dimension chains, elevation points and individual dimensions.

Linear dimensioning

The linear dimensioning function allows you to add dimension lines to horizontal, vertical and diagonal elements. First, you select, in any order, the points that need to be dimensioned. After you execute the function, the CAD program automatically generates the dimension lines and positions them in the desired location with the matching arrowheads, numbers and, if necessary, text. › Fig. 26, left These parameters can be set and subsequently edited in a dialogue box. Dimension points can also be entered into or deleted from existing dimension chains. The software independently computes all new distances.

Elevation dimensioning

Elevations in views and sections are also defined by elevation points. These are represented by an equilateral triangle and displayed directly in the drawing along with numerals. This dimensioning method is intended to measure and display height differences only. › Fig. 26, right The important points are simply clicked with the mouse, and a reference elevation is selected. Afterward the CAD program calculates the elevation and displays it at the chosen spot.

DESIGN TOOLS

The design tools introduced in this chapter are stand-ins for the large number of available options in CAD programs, where they depend on the individual software configurations. They are not necessarily identical in these different CAD programs, nor do they work in the same way. You will need to learn the special functions of the CAD program you choose and familiarize yourself with its special features. Ultimately, choosing a suitable CAD system depends not only on product quality but also on your own particular way of drawing.

Grids

Grids play an important role in the design of structures. They bring clarity to the work, particularly when large objects are involved, and provide critical support for the design process. For instance, a support grid can simplify the process of designing a building's loadbearing structure, and an axis grid can make it easier to design façades. Grids can be used for these same purposes in CAD and are a constructive drawing tool as well. ⟩ Fig. 27

Snap functions

To create accurate drawings, you must draw objects using precisely defined points. Using coordinates ⟩ Chapter Coordinate systems you can enter these points via the keyboard. Even so, the mouse is a much quicker and more intuitive entry tool for CAD drawings. The snap function, accessed via the menu bar or special toolboxes, allows you to select a point precisely with the virtual drawing and marking tool and prevents you from erring by a few millimetres within the drawing. It "snaps" onto drawing elements – that is, it selects them with great precision. The above-mentioned grid points can also serve as snap points: if you move the drawing tool close to one of these points on the screen, the CAD software marks it with a symbol. If you then execute the LINE command, specifying the starting point of a line with a mouse click, it will become the snap point. ⟩ Fig. 28 This ensures the necessary precision for drawing.

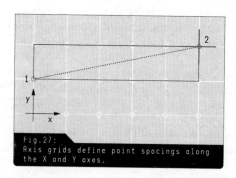

Fig. 27:
Axis grids define point spacings along the X and Y axes.

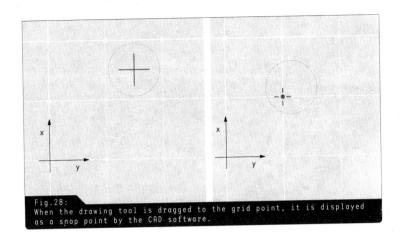

Fig. 28:
When the drawing tool is dragged to the grid point, it is displayed as a snap point by the CAD software.

The snap function usually works only in conjunction with other drawing commands. The required command must first be selected if you want the CAD software to activate the snap points automatically once they are within the drawing tool's snap radius.

Object snap

Snap points may be used in various ways. They may incorporate the end and mid points of lines, the corners of squares, the centres of circles, the intersection points of two objects and so on. › Figs 29–31 A symbol, displayed temporarily, illustrates the proximity of a potential snap point to the drawing tool – similar to the process of snapping onto grid points. The snap radius defines the maximum distance within which the drawing tool activates the snap point. Hence, the required snap point is temporarily activated once it is enclosed by an imaginary circle around a specific point whose radius usually measures a few millimetres on the screen.

\\Tip:
A grid consists of points, at user-defined distances, which can be connected with lines. For instance, a grid can be formed by the same sized spaces corresponding to 1 m on the X and Y axes. In CAD, grids can be arbitrarily moved within a drawing, just like the zero point of the underlying user-defined coordinate system (UCS).

\\Hint:
Precision is crucial when drawing with CAD, since coordinates and drawing elements are usually linked, and even small mistakes made at the beginning of a drawing can quickly become magnified as the work continues.

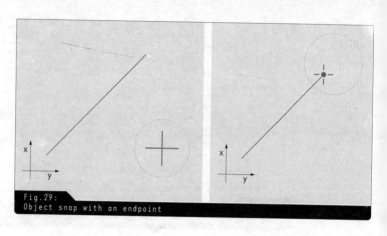

Fig.29:
Object snap with an endpoint

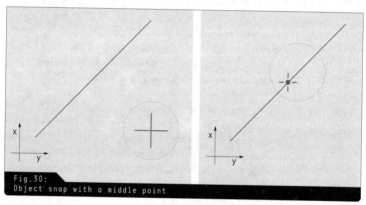

Fig.30:
Object snap with a middle point

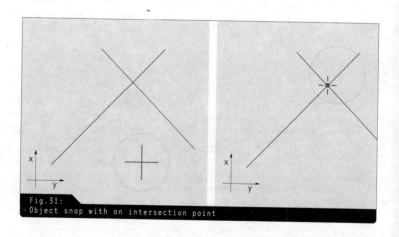

Fig.31:
Object snap with an intersection point

Angle specifications

Precise angle specifications are one of the most important criteria for creating precise technical drawings. CAD programs create precise angles in different ways. They usually allow you to specify the desired angle in a dialogue box via the keyboard or an interactive angle-degree selector. In many CAD programs, you are also given the option of an angle snap command with predefined angle increments if you hold down the shift key while drawing. Additional tool functions can normally be used as drawing aids and are available in toolboxes. › Figs 32–35

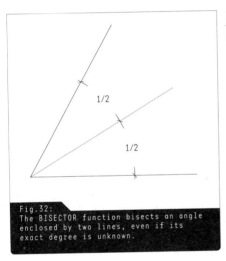

Fig.32:
The BISECTOR function bisects an angle enclosed by two lines, even if its exact degree is unknown.

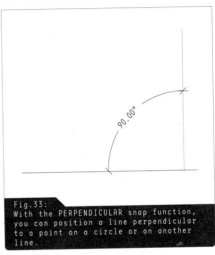

Fig.33:
With the PERPENDICULAR snap function, you can position a line perpendicular to a point on a circle or on another line.

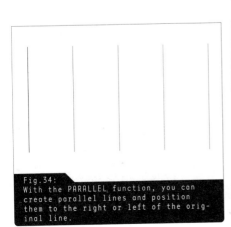

Fig.34:
With the PARALLEL function, you can create parallel lines and position them to the right or left of the original line.

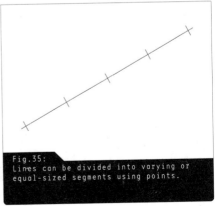

Fig.35:
Lines can be divided into varying or equal-sized segments using points.

MODIFICATIONS

One of the great advantages of CAD is that it allows you to edit and change objects after they have been drawn. This means that, in the design process, you can execute modifications and optimizations with relatively little effort. You can also, if necessary, modify templates and adjust them to meet a variety of requirements. There is no need to recreate every drawing from scratch, and it is possible to streamline the generation of standard objects. › Chapter Program libraries

Associativity

CAD elements can be "associatively" modified. For instance, if you fill a square with hatching and subsequently change its shape, the hatching automatically follows its new perimeter. › Fig. 36 The square and its filling are linked and geometrically dependent on each other. Another example of associativity is a component that is altered in combination with dimensioning lines: if the length of the component is modified, the corresponding dimension chains automatically change too – provided there is an associative link between the two. This means that it is not necessary to create new dimension lines for an edited component that has already been dimensioned.

Grouping

In CAD, objects are often made up of various elements that form a unit in the drawing but are not necessarily linked. For instance, the drawing of a bed may consist of several lines, which must be individually selected if they are modified. Here, it is possible to group together several elements so that you do not need to activate all of them when you select an object. Depending on the CAD system used, they are combined to create GROUPS, BLOCKS or SEGMENTS. Defined as such, they form units that can be broken up again if necessary. This normally makes the work much easier.

Modifying points

The process of drawing objects becomes much more flexible because users are able to modify individual points or several points at the same time. Simple changes in line length can be made with the mouse, and sur-

\\Tip:
An existing building layout can easily be modified by selecting all the points on one side of the building and moving them in the desired direction within the UCS. All the selected points will move as well, and the lines and components will automatically be lengthened.

\\Hint:
When you make other types of modifications, it is also helpful to use a selection window to activate elements because this tool provides greater control in selecting and changing elements.

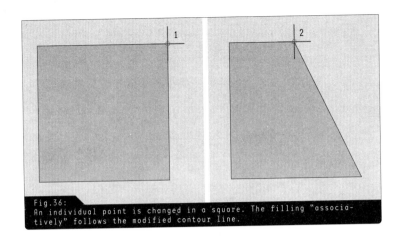

Fig. 36:
An individual point is changed in a square. The filling "associatively" follows the modified contour line.

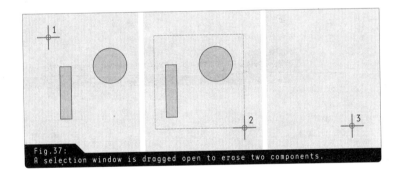

Fig. 37:
A selection window is dragged open to erase two components.

faces and three-dimensional objects can also be altered easily. > Chapter The third dimension

After activating the command, you can, for instance, grasp the corner of a rectangle using the snap function and move it to the desired position within the virtual drawing area. > Fig. 36

Erase

The erase function is executed using the selection tool. It can be used to remove any element in the drawing completely, or to erase several elements at once. After activating the erase function, you simply drag open a window with the mouse. Depending on the basic settings, you can erase all objects that are located within the selection window or that are intersected by it. > Fig. 37

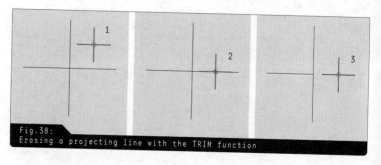

Fig.38:
Erasing a projecting line with the TRIM function

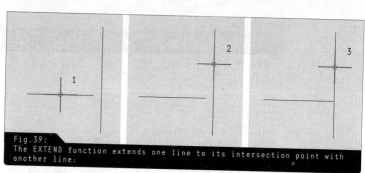

Fig.39:
The EXTEND function extends one line to its intersection point with another line.

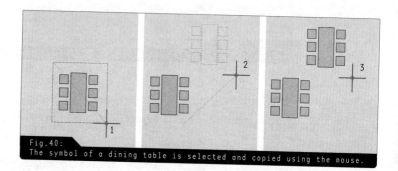

Fig.40:
The symbol of a dining table is selected and copied using the mouse.

\\Tipp:
You can usually use the TRIM and EXTEND functions for either a single line or for several lines. Hence, any number of lines can be selected, trimmed and extended at the same time.

\\Tip:
Snap points can also be used to grasp an object. They allow you to snap onto the object and move it to another location with great accuracy. Snap points can also be used to position it there.

Trimm

It is also possible to trim or remove parts of elements. With the trim function (also called SHORTEN or DELETE ELEMENT BETWEEN POINTS in some CAD systems), a line segment running between two points or extending beyond a single point can be erased by activating the function and clicking the line segment with the selection and drawing tool. › Fig. 38

Extend

The opposite command, EXTEND (also called LENGTHEN or CONNECT TWO ELEMENTS), extends a line to its intersection point with another line. › Fig. 39

Copy

The COPY function is one of the most important drawing commands and forms the basis of an efficient work process. Generally speaking, before redrawing an object, you should always consider whether a drawn object can first be copied and then changed. This is especially true of complex objects that are time-consuming to create, whether these are objects you draw yourself or library elements. › Chapter Program libraries

To copy an object, first select it and then drag it to the desired location by holding down the selection and drawing tool. The original object will remain in its initial position. › Fig. 40

Mirror

You can use the MIRROR function to create mirror-images of objects along any axis you choose. You can also create a copy – which is naturally inverted – and continue to use the original. After activating the command and selecting the object, you specify a mirror axis at the desired distance and angle to the original and then mirror the object across the chosen axis. You can also use lines and edges of existing elements as mirror axes.

In combination with the copy function, this command makes it possible to draw symmetrical objects efficiently: only one half must be drawn, and the other half can be mirrored across the middle axis. › Fig. 41

> \\Example:
> If you are drawing the floor plan of a twin house, you can mirror either all or a part of the layout across the middle axis before continuing with your work. The same approach lends itself well to drawing façades and other building components quickly and efficiently.

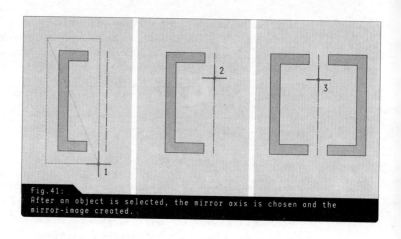

Fig.41:
After an object is selected, the mirror axis is chosen and the mirror-image created.

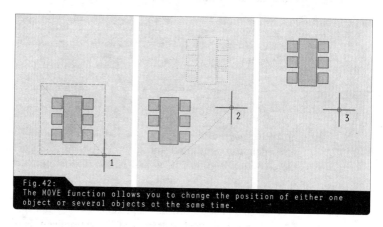

Fig.42:
The MOVE function allows you to change the position of either one object or several objects at the same time.

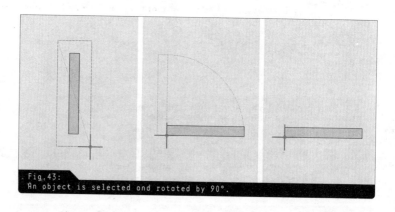

Fig.43:
An object is selected and rotated by 90°.

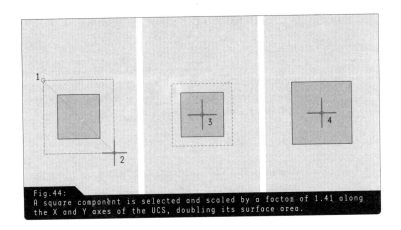

Fig. 44:
A square component is selected and scaled by a factor of 1.41 along the X and Y axes of the UCS, doubling its surface area.

Move

The MOVE function is quite similar to the copy function. The only difference is that the object is moved and no copy is made. › Fig. 42

Rotate

With the ROTATE function, drawn objects can be rotated an arbitrary distance around an arbitrarily selected point. If all that is needed is to change the object's orientation, its middle point represents an ideal snap point. With the mouse, you specify the original angle – normally the positive X axis in the UCS. Starting here, the object is then rotated around its centre. Alternatively, you can use the keyboard to enter the rotation angle after determining its pivot point. This method can be employed to rotate a vertical structure by 90°. › Fig. 43

Scale and stretch

The size, length and shape of all drawn objects can be changed with great accuracy. The SCALE or STRETCH function allows you to scale an object in all directions of the coordinate system. › Fig. 44

THE THIRD DIMENSION

The preceding chapters dealt primarily with drawing functions on the construction plane, which, from a geometric perspective, is the equivalent of a two-dimensional sheet of drawing paper. One great advantage of CAD systems is that they allow you to work in space, that is, in three dimensions.

Whereas a drawing only reproduces a view of a 2D structure, three-dimensional design in CAD allows the designer to give shape to the 3D object. A <u>Z axis</u> must be added to the basic CAD reference system – the coordinate system – in order to transform a two-dimensional plane into three-dimensional space. The most commonly used coordinate system is the Cartesian system discussed above, which is merely expanded to include the Z axis. The Z axis rises perpendicularly from the zero point of the plane defined by the X and Y axes and makes it possible to define the point P(3,5) from Fig. 5, page 16, in three-dimensional space. › Fig. 45

THREE-DIMENSIONAL DESIGN

In the third dimension, the two-dimensional construction plane becomes an element of the three-dimensional workspace. However, as a rule,

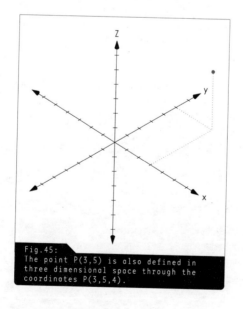

Fig.45:
The point P(3,5) is also defined in three dimensional space through the coordinates P(3,5,4).

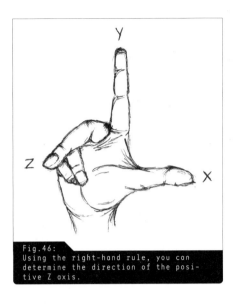

Fig.46:
Using the right-hand rule, you can determine the direction of the positive Z axis.

use of this workspace normally relies on two-dimensional pointing and output devices such as a mouse and a screen. This can take some getting used to when complicated, three-dimensional drawings are involved.

Projections

Whereas two-dimensional drawings generally only involve the projection of a plane (e.g. floor plan or view), three-dimensional representations allow you to alternate between the top or bottom view, the four different elevations and the isometric representations. › Fig. 47 All these presentation methods create projections of the same three-dimensional objects and

\\Tip:
The "right-hand-rule" can help you get your bearings within the Cartesian coordinate system. If the directions of the X and Y axes in a three-dimensional coordinate system are known, this rule shows the direction of the positive Z axis. If the thumb and index finger of your right hand point in the positive X and Y directions, your extended middle finger represents the positive Z axis (see Fig. 46).

\\Tip:
When designing, you should start thinking and working in three dimensions as early as possible, since many of the links and relationships within a structure are best checked in spatial terms. Along with simple cardboard models, abstract virtual 3D mass models are particularly useful, since they can be constructed relatively easily and provide quick results (see Chapters Construction methods and Visualization).

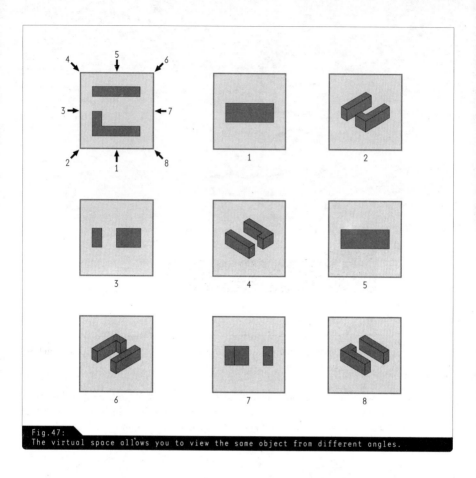

Fig. 47:
The virtual space allows you to view the same object from different angles.

facilitate both clear spatial presentation in the respective viewing mode and the editing of particular points of a 3D design. Most CAD programs allow you to view several projections simultaneously in different windows on the screen. You are thus able to see the object from all directions.

Models   Three-dimensionally defined objects have different qualities and, depending on the design method used, can be categorized as volume models, plane models or edge models.

Volume models   A volume body is a solid object that contains a great deal of information within the CAD system. › Fig. 48 As well as providing volumes and the information derived from them such as mass and centre of gravity, volume

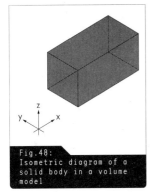

Fig.48:
Isometric diagram of a solid body in a volume model

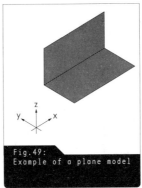

Fig.49:
Example of a plane model

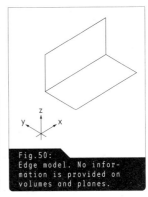

Fig.50:
Edge model. No information is provided on volumes and planes.

models can be used to define surface characteristics and specific materials. › Chapter Visualization This enables the production of a virtual image of a structural component that exhibits the same characteristics under programmable virtual conditions as its real counterpart. Moreover, the image allows for interpolations, which can be geometrically defined within the volume. This makes it possible for bodies to be geometrically combined, subtracted from one another or positioned to form intersections. › Chapter Modelling

Plane models

As their name suggests, plane models consist exclusively of planes and do not have any defined volume. In the CAD program planes can be defined in terms of the visual characteristics of their surfaces, and they can be rotated and inclined at any angle within the virtual workspace. › Fig. 49

Edge model

An edge model uses three-dimensional lines to simulate only the edges of a body. No information is provided concerning surfaces or volumes,

> \\Hint:
> Plane models can enclose a volume without the latter being geometrically defined like a volume model. This produces objects whose shell provides information on surface characteristics and may resemble a volume model. Nevertheless, this shell consists exclusively of two-dimensional planes without material thickness. The object is just a hollow body.

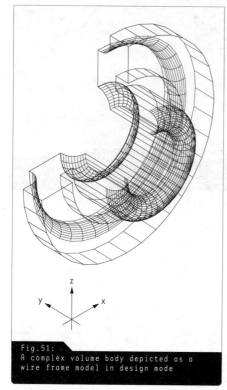

Fig.51:
A complex volume body depicted as a wire frame model in design mode

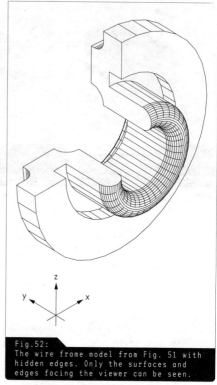

Fig.52:
The wire frame model from Fig. 51 with hidden edges. Only the surfaces and edges facing the viewer can be seen.

which are thus not geometrically defined. Edge models play only a minor role in architectural CAD since they are highly abstract representations.
> Fig. 50

Wire frame model

In the 3D design mode of most CAD programs it is standard for all spatial objects to be depicted as wire frame models in order to take all point coordinates into account.

A wire frame model – which should not be confused with the edge model described above – is a skeletal depiction of a 3D object using three-dimensionally defined lines. In the design mode, both volume and plane models are shown as wire frame models, although in this special type of constructive representation information about volumes and planes is not included. In the virtual workspace a wire frame representation can be observed from any chosen point, allowing you to check spatial relationships such as the distances between structural components and overlaps. Some practice is required to be able to differentiate between "front" and "back" in a complex wire frame model, because the model shows spatial depth even though a two-dimensional representation appears on the monitor. › Fig. 51 Some CAD programs function without the depiction of wire frame models and visualize three-dimensional objects directly with shaded or coloured surfaces.

Hidden edges

If required, all edges located behind three-dimensional objects and planes can be concealed to provide better orientation within the drawing. This makes the representation much more comprehensible and is a preliminary stage in computer rendering, although in contrast to a completed visualization it can still be modified. › Fig. 52 and Chapter Visualization Other CAD programs give the command for this representation option as HIDE or HIDDEN LINE.

Generating floor plans, views and sections

Technical drawings typically depict floor plans, views and sections of entire structures or structural components as detail drawings. However, in three-dimensional design mode you can produce complete virtual models that, if necessary, can provide a basis for generating the two-dimensional figures of a technical drawing.

Floor plans and horizontal sections

In default mode, the construction plane of a CAD system is represented as a top view, i.e. the user looks down on the XY plane of the UCS and thus onto a virtual model. In order to represent a particular floor of a virtual building as a floor plan, a horizontal section plane is generated

\\Tip:
Calculating the dimensions of hidden lines can be time-consuming. Where the scale and detail of a 3D representation has been clearly specified prior to the construction of the 3D model, you should, where applicable, avoid irrelevant elements and unnecessary depth of detail. If you require a 3D model for both a rough representation and a detailed depiction, details such as doorknobs should be represented on an additional layer so that they can be excluded from the calculations relevant to the rough model (see Chapters Transparent planes – the layering principle and Visualization).

\\Hint:
Along with a design mode window, many CAD programs also offer an optional 3D depiction. This can be activated as an additional window and can show three-dimensional objects with hidden edges and shaded or coloured surfaces. Objects depicted in this way can usually be rotated and examined from all sides.

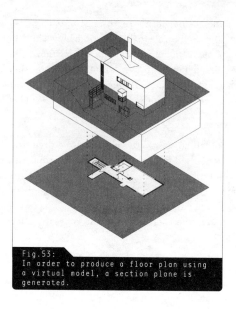

Fig.53:
In order to produce a floor plan using a virtual model, a section plane is generated.

which slices through all walls and objects within the building at a particular height and thus depicts them as elements of a floor plan. › Fig. 53

Views and vertical sections

Most CAD programs employ the same tool functions to generate views and sections. In technical drawing, a section is also a view: one that cuts through a building or component, thus allowing you to see the interior of the object. In contrast to the generation of a floor plan, longitudinal and lateral cuts produce vertical sections, which slice through an object in a

\\Hint:
The drawings in Figs 53–55 are based on a design by Le Corbusier for an apartment block in Vaucresson (1922), although the dimensions and details do not correspond exactly to the original.

\\Tip:
A horizontal section cuts through objects parallel to the XY plane of the UCS. It follows that a vertical section bisects objects parallel to the Z axis. Some CAD programs provide predefined horizontal section planes as floor levels (e.g. Graphisoft ArchiCAD), and you can alternate between these while working on the different floor plans of a building (see Chapter Transparent planes – the layering principle).

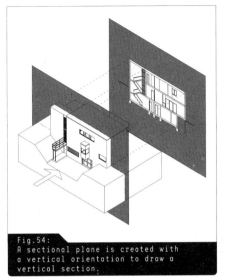 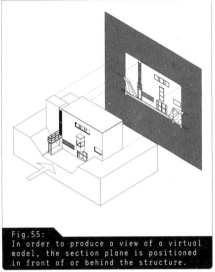

Fig.54:
A sectional plane is created with a vertical orientation to draw a vertical section.

Fig.55:
In order to produce a view of a virtual model, the section plane is positioned in front of or behind the structure.

longitudinal or lateral direction. The first step is to establish the length as well as the line and depth of vision of the section plane. (In the case of a view, this section plane is located in front of or behind an object.) Within the top view, the CAD software then generates the symbol for a section boundary, which you can select to access a depiction of the defined section or view. Depending on the CAD system used, this depiction is shown in a special section-view window or is generated as a two-dimensional drawing, which you can then work on. > Figs 54 and 55

\\Hint:
Some programs allow you to create and label a view or section and assign it a separate window even though the relevant virtual model is still linked with 3D data (e.g. Graphisoft ArchiCAD). As a result, when you work on elements in construction mode, any changes you make are automatically reflected associatively in the projection. Alternatively, the projection can be generated on an additional layer consisting only of two-dimensional lines.

CONSTRUCTION METHODS

Three-dimensional construction methods function similarly to two-dimensional drawing commands. That said, there is a fundamental difference: coordinates are not only defined via the X and Y axes but are also positioned in space using the Z axis. › Chapter Three-dimensional design

Points, lines and planes in space

As when drawing points, lines and planes in the construction plane of the UCS, you use a drawing function to draw them in three-dimensional space. Working within a spatial projection takes some getting used to, since you are often unable to see the real lengths of spatial representations, because not all objects are shown parallel to the coordinate axes and thus appear shortened or elongated in different projections. › Fig. 56

Cuboids

Simple bodies can be generated as volume models with the drawing functions. For example, a cuboid is initially defined by its base or base area by using its diagonal and then assigned a height. › Fig. 57 The base is always positioned on the construction plane and is parallel to the XY plane of the UCS.

Cylinders

A cylinder can have either a circular or an elliptical base. Once this has been defined, the height of the cylinder is determined using the Z axis.
› Fig. 58

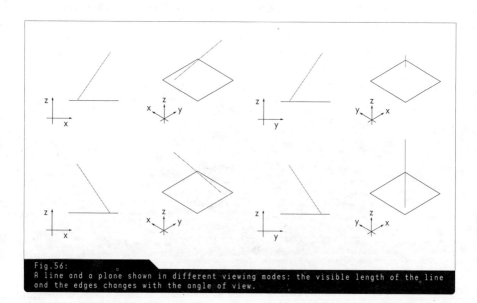

Fig. 56:
A line and a plane shown in different viewing modes: the visible length of the line and the edges changes with the angle of view.

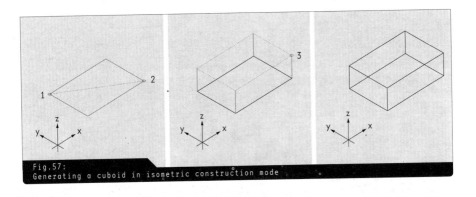

Fig.57:
Generating a cuboid in isometric construction mode

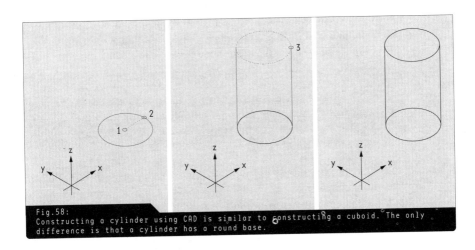

Fig.58:
Constructing a cylinder using CAD is similar to constructing a cuboid. The only difference is that a cylinder has a round base.

\\Tip:
In three-dimensional space, planes can be generated as polygon planes. In this case, you can use not only simple horizontal and vertical orientations but also any angle of inclination you require. If a plane is supposed to have a certain inclination, the simplest method is first to generate it in two dimensions in the top view and then to rotate it in three dimensions.

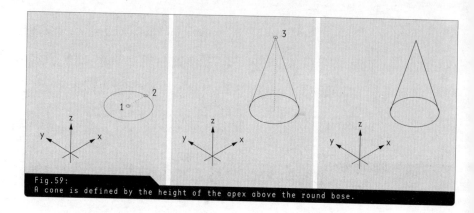

Fig.59:
A cone is defined by the height of the apex above the round base.

Cones

A cone is defined by a circular or elliptical base and narrows vertically to a point. When generating a cone, first define the base in terms of the base area and then determine the apex as a point using the Z axis.
› Fig. 59

Spheres

In CAD a sphere is defined using a midpoint in the base and a radius or diameter. The sphere's lines of latitude are positioned parallel to the XY plane of the UCS, and the vertical middle axis is congruent with the Z axis. › Fig. 60

Extruded volume bodies and planes

The 3D drawing functions described above make it very easy to generate simple geometric bodies with a single drawing command. Not all CAD programs offer these functions and in such cases the possibilities of geometric design are limited by preset form definitions. Another method of generating volume bodies is <u>extrusion</u>, which involves raising the body out of the ground plan sketch. This function allows you to transform two-dimensional objects consisting of lines, polylines and splines into volume bodies and planes.

This method is useful above all when working on objects that include rounded edges, bevels and other geometric peculiarities, since they are difficult to define without a preset basic shape. It is important that the figure consisting of line segments is closed and does not overlap with other figures. If the basic form is not closed, a plane is generated rather than a volume body.

Different CAD programs use different terms for extrusion, such as RAISE, TRANSLATION and TRAJECTION. Extrusion is carried out along a construction line or through entry of a height value. In the process of extrusion, the

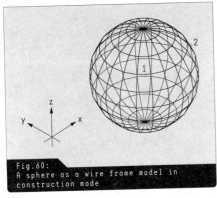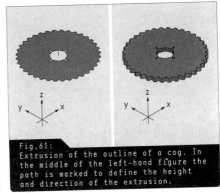

Fig. 60:
A sphere as a wire frame model in construction mode

Fig. 61:
Extrusion of the outline of a cog. In the middle of the left-hand figure the path is marked to define the height and direction of the extrusion.

two-dimensional basic shape that is to be transformed into a volume body is usually shown as an OUTLINE. The construction line along which the outline is to be extruded is often labelled PATH. Some CAD programs require both the basic shape and the three-dimensionally defined path in order to carry out the extrusion operation (e.g. Nemetschek Allplan). Others (e.g. Autodesk Architectural Desktop) need only a basic shape and a particular height, width and length entered as coordinates along the coordinate system axis in order to extrude a model from the outline. ⟩ Fig. 61

Rotational solids

The ROTATION command enables you to rotate and generate a two-dimensional basic shape around the X or Y axis of the underlying UCS – or around any other axis defined by two points. As with the EXTRUSION command, ROTATION is particularly useful when dealing with objects with rounded edges and other details that are difficult or impossible to form using simple 3D drawing commands. Here, too, the outline of the underlying profile must be closed and cannot have any lines overlapping. In the

\\Tip:
Extruded volume bodies are very useful when generating terrain models. As in a cardboard model, where the terrain profile is reproduced in different layers, you can construct similar layers on the computer for every basic shape and extruded height and place them on top of one another.
Further information on manual modelbuilding can be found in *Basics Modelbuilding* by Alexander Schilling, Birkhäuser Publishers, Basel 2007.

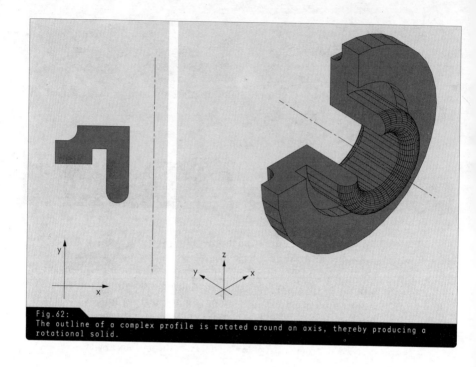

Fig.62:
The outline of a complex profile is rotated around an axis, thereby producing a rotational solid.

example provided in Figure 62, the profile positioned in the XY plane is rotated around an axis that is parallel to the Y axis of the UCS. The process is comparable to using a potting wheel that is spinning around a vertical axis, where the rotation makes it possible to form the modelling material to any shape you wish.

MODELLING

The previous chapters have presented construction methods that involve using a drawing function or deriving an outline to generate simple and complex volume bodies. Volume bodies can also be modelled, i.e. their fundamental shape can be altered. Depending on the geometry involved, one or more points of a body can be modified and moved in the directions prescribed by the axes of the UCS. For example, in only a few steps you can shape a cuboid into a wedge. › Fig. 63 Or – if you need to tackle a more complex operation – you can change the overall height of a building.

However, modelling is not restricted to simple manipulations of points. It can also be carried out in a large number of ways using com-

Construction with volume bodies

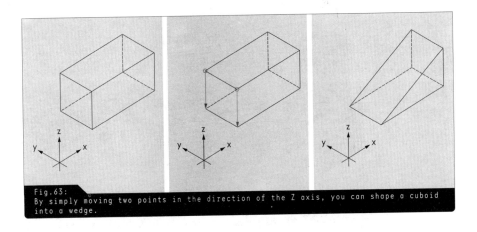

Fig. 63:
By simply moving two points in the direction of the Z axis, you can shape a cuboid into a wedge.

mands that are at times more complex and at others quite simple. For example, the fact that volume bodies can be added to and subtracted from one another opens up a wide range of options. As with many drawing functions, these geometric operations go by different names in the various software packages, e.g. COMBINATION, CSG (short for Constructive Solid Geometry) and BOOLEAN OPERATIONS. The basic functions of these construction methods are based on addition, subtraction and intersection.

Addition

Addition allows you to combine two or more volume bodies into a single object. Here, the order in which the objects are selected for addition is unimportant. After they have been combined they can be selected and worked on only as a group, since they now constitute a newly defined object. This operation is particularly useful when separate building components need to be combined into a single structure. > Fig. 64b

Subtraction

Subtraction allows you to separate bodies from one another and to form different objects. In this operation the order of object activation is decisive, since only the first object to be activated remains intact while the second is erased along with the intersection volume. The second object thus determines the negative shape of the interpolation in the first. > Fig. 64c This operation makes it very easy, for example, to generate recesses for windows in wall components.

Intersections

When generating intersections, the intersection volume of two bodies is formed as a new object. The form of the object to be generated is determined by the way in which the two bodies are interpolated and is then shown as a positive shape. Areas that do not overlap are removed. > Fig. 64d

53

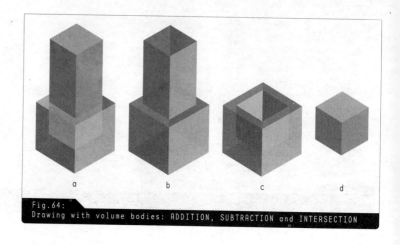

Fig. 64:
Drawing with volume bodies: ADDITION, SUBTRACTION and INTERSECTION

ARCHITECTURAL ELEMENTS

Generally, CAD elements are defined geometrically or via the coordinates of their points. This geometric information allows the calculation of lengths, widths and heights and thus determines planes and volumes. In addition, it is possible to link this information to other specifications and to define general and specific characteristics. CAD programs that specialize in architectural drawing and visualization employ these associative possibilities for the efficient generation of typical architectural components. As a result, complex building components such as multi-layered walls, windows and stairs can largely be predefined and thus drawn with optimum accuracy. These components therefore no longer need to be constructed in many individual steps and subsequently combined; instead, they are predefined in a dialogue box. Moreover, depiction parameters for pens, hatching and the surfaces of a building component can be defined in terms of colour and transparency. › Chapter Visualization Furthermore, it is possible to prepare and categorize different building components for the subsequent calculation of planes and volumes. › Chapter TAI In this way an entire building can be efficiently constructed, visualized and quantified in virtual form.

Drawing architectural elements

› 📏

Similarly to two-dimensional objects, architectural elements are usually constructed in the ground plan from the top view. The fundamental difference from 2D drawing is that you draw not simply a single line, but the entire component, creating several layers to record structure, height and the parameters for representing materials.

Walls

If you are drawing a wall as a complex component with a WALL tool, a dialogue box linked to the drawing function allows you to define the

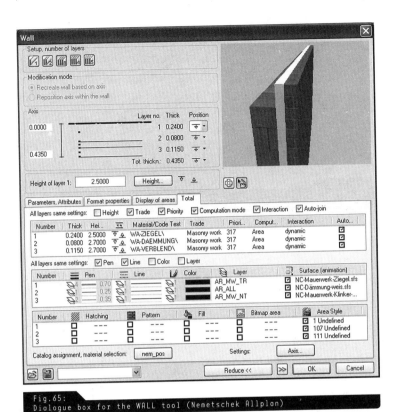

Fig. 65:
Dialogue box for the WALL tool (Nemetschek Allplan)

characteristics of this component prior to the actual drawing process.
> Fig. 65

This means that you can set the default setting for wall height and structure to generate single or multi-layered elements. The layers are defined as lines with different pens and, if necessary, given additional at-

\\Hint:
An advantage of this type of design is the capacity to combine certain building components that associatively influence one another. For example, the material connections between two overlapping, multi-layered walls form automatically and do not require laborious design work.

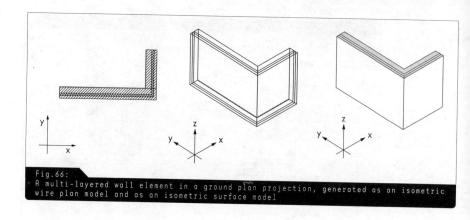

Fig. 66:
A multi-layered wall element in a ground plan projection, generated as an isometric wire plan model and as an isometric surface model

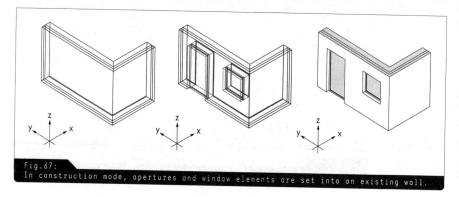

Fig. 67:
In construction mode, apertures and window elements are set into an existing wall.

tributes that can be used for subsequent visualization or calculation of quantities. When you now draw the wall, the depiction will include all predefined information and the result will be a complex component. › Fig. 66 In this way a whole series of steps which are required when drawing 2D or even simple 3D elements are combined in a single drawing function.

Apertures, windows and doors

› 🛈

After you have drawn the walls as components, you can generate windows and doors in the wall planes in any format you choose. Rather than merely defining the width and height of an aperture, you can also insert and fit windows and doors as finished components with frames, partitions and panels. › Fig. 67

Ceilings

As in the automatic generation of walls, ceilings are defined geometrically in terms of shape, panel thickness and position, and are predefined in the depiction by means of hatching and surface characteristics.

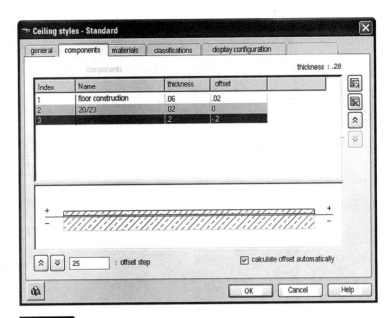

Fig. 68:
Dialogue box for CEILING tool (Autodesk Architectural Desktop)

If required, other attributes relating to the calculation of quantities and visualization can also be indicated. › Fig. 68

Once you have predefined the ceiling in this way, you can draw it with a base area of your choosing using a polyline and portray it as a completed component. Ceiling recesses such as stair openings can be constructed subsequently as polylines and recessed into the ceiling plane.

Stairs

When drawn by hand, stairs require laborious construction and calculation. CAD software combines the height and width of the stairs with

> \\Hint:
> Wall openings and inserted components such as doors and windows are associatively linked. The building components are automatically adapted to the structure of the wall with an appropriate embrasure.

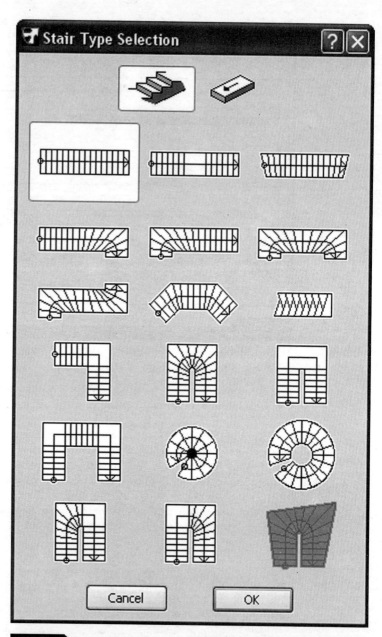

Fig.69:
Dialogue box for STAIR tool: different stair types can be selected (Graphisoft ArchiCAD).

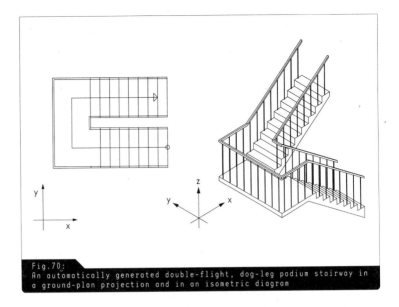

Fig. 70:
An automatically generated double-flight, dog-leg podium stairway in a ground-plan projection and in an isometric diagram

the stair ground plan and the height between floors to construct the stair geometry, all of which parameters can be altered during or even after the construction phase. Once you have selected the shape of the stairs, a dialogue box allows you to select and set their geometry and construction details. ˃ Fig. 69

The stairway and all its construction characteristics are automatically calculated and generated. They can then be checked in the three-dimensional view and if necessary altered again. ˃ Fig. 70

\\Hint:
Some of the CAD programs that provide a stair tool initially predefine the stairway as a complete component in a dialogue box and allow you to save it as a library element before you place it within the virtual drawing area (e.g. Graphisoft ArchiCAD.) Other programs use a predefined 2D ground plan as their basis, to which you add geometric, constructive characteristics by selecting polygonal points (e.g. Nemetschek Allplan).

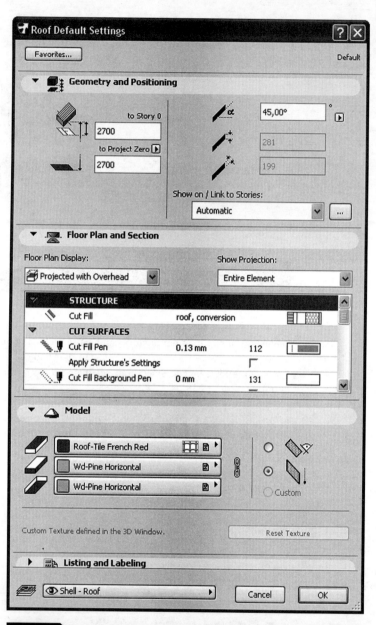

Fig.71:
Dialogue box for ROOF tool: along with line attributes, parameters can also be defined for storey height, roof inclination etc. (Graphisoft ArchiCAD).

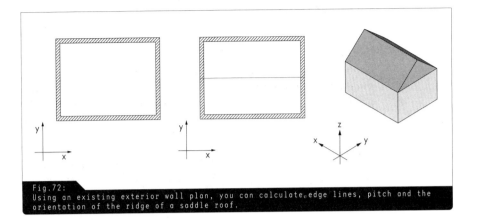

Fig.72:
Using an existing exterior wall plan, you can calculate edge lines, pitch and the orientation of the ridge of a saddle roof.

Roofs

Roofs are geometrically determined by the shape of the base area, the roof profile and inclination, the ridge, the edge height, and in some cases by the overhang. In many CAD systems, simple roofs can be designed with a ROOF tool by drawing the base area of the roof as a polyline and entering all desired parameters of the roof profile using the corresponding dialogue box. > Figs 71 and 72

> \\Tip:
> Once the shape of the roof has been generated, many CAD programs allow you almost automatically to generate a rafter plan and definitions of the roof structure. These programs are thus able to take a virtual model from the design stage to the construction planning stage in only a few steps.

VISUALIZATION

In architecture, the term visualization refers to the pictorial representation of a detail or a planned building. One of the advantages of computer simulation over technical drawing as a means of visualization is that computer imaging is easier for non-professionals – including many building clients – to comprehend, because it provides a visual translation of the technical information contained in a construction drawing.

Visualization – a design tool

Computer visualizations serve different purposes. Planners can use them as a design tool to check the proportions of structural elements and spatial relationships between components and their surroundings.

Fig.73:
Visualization as a design tool: cubatures, lighting and materials can be examined in different phases of the design process.

In addition, visualizations can be used to evaluate the effects of various materials and lighting conditions when planning building projects. Using simple and relatively abstract virtual models at an early stage in the design process, you can check many aspects of the design, such as the suitability of materials and structural proportions. You can thus generate detailed computer simulations that, depending on the stage of the design and the degree of differentiation incorporated into the original virtual model, can help you make decisions during the planning process. The possibilities open to you range from simple mass models to virtual buildings, which reflect the current stage of planning. > Fig. 73

Computer visualizations also enable you to evaluate the advantages and disadvantages of different design approaches in the context of the planned location. While the exterior space remains unchanged, you can examine the relationship between the different architectural languages of alternative structures and the location. > Fig. 74

> Visualization for presentation

Architectural visualizations – like other methods of representation such as drawings and scale models – can also be used for the final presentation of a design. Depending on the requirements of the specific design and the target group, a visualization can be photorealistic or abstract. A successful presentation need not necessarily use photorealistic depictions: simple cubatures and abstracted surfaces can be non-realistic yet still communicate many aspects of the envisioned architectural effect. > Fig. 75

If possible, virtual models should be constructed to reflect subsequent visualizations. It is an advantage to adapt the degree of detail in a model to the possibilities of the visual presentation in order to use time efficiently in both the design and computation of a visualization. > Chapters

\\Hint:
In order to generate a visualization you need appropriate software, called rendering software (see Chapter Rendering parameters and Appendix, Table 4). A number of CAD programs include an integrated rendering module that executes at least the initial steps in the visualization process. To achieve good, detailed results, you will often need special software, referred to as a renderer.

\\Tip:
Many rendering programs allow you to position a pixel image in the background of a virtual object and thus to create a real backdrop in a virtual situation. For example, by using a digital photo of surrounding structures as a background image, you can depict a planned building in its actual future surroundings.

Fig. 74:
Different architectural approaches in the same urban-planning context

Three-dimensional design and Rendering parameters For instance, details such as protuberances, recesses, window allocations and the exact differentiation of materials are not necessary for the simulated aerial view of an urban design, because they cannot be seen from a great distance. ﹥ Fig. 76

It is only at closer distances that details of the building cubatures and the structure of materials become visible and thereby important for the visualization. Features of the external space such as trees, cars and people can also fill a visualization with life, creating – as in a photo –

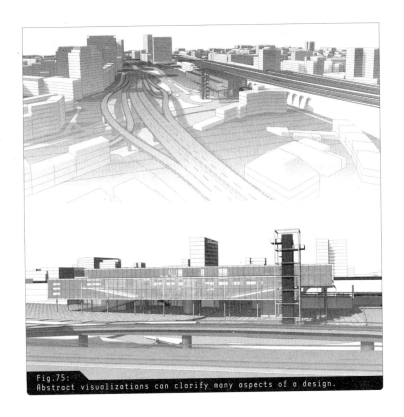
Fig.75:
Abstract visualizations can clarify many aspects of a design.

Fig.76:
Visualization of a planned urban structure

Fig. 77:
The details of a building and its surrounding become more important as the viewer draws closer.

familiar references to everyday life and providing the viewer with a means to grasp the scale of the objects being depicted. › Fig. 77

Interior space also plays an important role in architectural visualizations. When creating a realistic interior visualization, you should pay attention to structural details and surface qualities, since interior views are often presented in close-up. › Fig. 78

SURFACES

An important part of a computer visualization is the optical effect of surfaces, which are given their material quality by colours, textures and lighting effects.

> \\Tip:
> In a visualization of a high technical quality, virtual worlds seem perfect – and precisely for this reason unrealistic. The real world contains many "mistakes" that are significant for a computer visualization. Creating virtual façades with slightly uneven surfaces and colour nuances, for example, produces a better visualization with a much more realistic effect.

Fig.78:
Different depictions of an interior view in the visualization process

Colours

In the real world, it is the colour spectrum that allows us to perceive surfaces visually. The light that hits a surface is absorbed or reflected. If the entire colour spectrum is reflected by an object it appears white; if the entire colour spectrum is absorbed it appears black. Thus, our perception

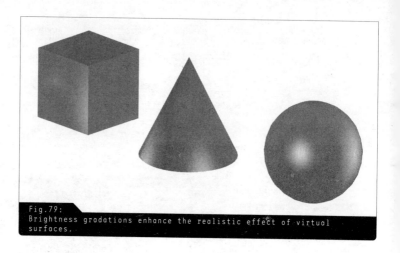

Fig.79:
Brightness gradations enhance the realistic effect of virtual surfaces.

of a colour depends on whether it is reflected or absorbed by the surface. For example, a blue surface reflects the colour blue while absorbing all other colour values.

An important method for creating realistic simulations of surfaces is selectively to change their colours. To have a realistic effect, the colour values of objects need to show nuances that are produced by differences in the way they reflect light. > Fig. 79

The sides that are turned away from the light source appear darker than those directly exposed to it. Moreover, highlights on strongly reflective surfaces appear white, irrespective of the colour of the object. As a rule, the possibilities for defining the rendered objects offered by the rendering software allow you to introduce variations in both the surface colour and the reflected colour, as well as in specific mirroring and transparency characteristics. If required, these can be set as percentages in a dialogue box.

You can use the electronic pen with which you drew the object to define its colour properties and reflective and transparency characteristics. Within the CAD system, the pen, or pen colour, is associated with the corresponding virtual information that is represented in the computer visualization as visible characteristics. > **Chapter The virtual drawing board**

Textures

Textures can be applied to the surfaces of a virtual object in order to represent a wide range of materials. This usually involves using a detail of

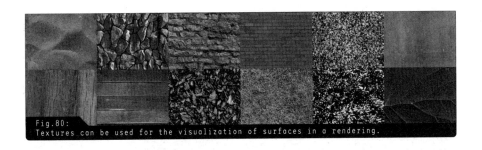

Fig.80:
Textures can be used for the visualization of surfaces in a rendering.

a <u>pixel image</u> – e.g. a photographic close-up of a brick wall. › Chapter Rendering parameters With the help of the so-called MAPPING function, you can place this pixel image on the virtual surface and adapt the textured area to the size you require. Finally, when the view of the texture – in this case, a virtual brick wall – is rendered, it has the same surface of its real correlate. This is how a piece of the real world is imported via a photo to the virtual model and used for the visualization. › Fig. 80

LIGHT AND SHADE

Well-placed light sources and selectively applied shadow effects are important for creating realistic surfaces and a particular atmosphere. Rendering software enables you to position and adjust different virtual light sources to simulate lighting conditions in exterior and interior space.

Virtual light sources

Virtual light sources can be used with differing intensities and colours. Moreover, it is possible to make light visible that under normal conditions is invisible until it falls on a surface – an effect that is similar to the light from a torch shining through fog.

\\Example:
While a normal window glass is transparent, it also reflects the surroundings to a certain degree. As a rule, you can check the effect of the settings for the respective surface materials in a renderer preview. Given a little practice with the software, you will thus be able to carry out the initial stages of simulating the object you have in mind.

\\Tip:
As a rule, a selection of standardized textures is provided by the rendering software. This means that you can try out some initial textures without having to photograph the environment you are dealing with. In addition, you can find a large selection by running a search on the keyword "texture" on the Internet.

69

Fig.81:
Light sources have a great influence on the atmosphere of a visualization.

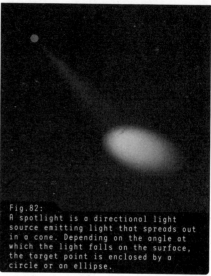

Fig.82:
A spotlight is a directional light source emitting light that spreads out in a cone. Depending on the angle at which the light falls on the surface, the target point is enclosed by a circle or an ellipse.

Fig.83:
A parallel light emits a directional shaft of light that does not form a cone.

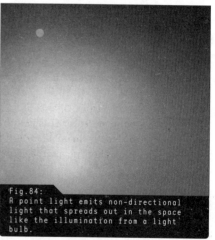

Fig.84:
A point light emits non-directional light that spreads out in the space like the illumination from a light bulb.

> 🗎 Spotlight, parallel light and point light are different virtual light sources that create distinctive light distributions and effects. › Figs 81–84

Ambient light   Apart from local light sources, ambient light can also be used in a design. This non-directional light is distributed equally throughout a space and constantly illuminates and colours the objects within this space. The brighter this light, the weaker the influence of other sources of illumination. You should use ambient light to fine-tune lighting and to avoid saturation or clouding in the visualization.

Intensity of illumination

The intensity of illumination is controlled via the colour of light, which is set individually for each light source in a scene. The brighter the colour, the stronger the luminosity of the light source. Pure white has the greatest luminosity. If you find that the luminosity of a light source is not sufficient, you can activate additional light sources. Since light colours can have all the colour values of the light colour spectrum, they can illuminate the space in every possible colour.

Shadows

Like real light sources, virtual light sources can produce shadows, thereby enhancing the realism of a virtual scene. Different methods of computing shadows can be used to produce hard, soft and area shadows.
› Figs 85–87

PERSPECTIVE AND VIRTUAL CAMERA

The choice of the right perspective is key to computer visualization. › Figs 73-78 All factors relevant to the image should be taken up in an appealing image format, if necessary with striking details. Image content should also be shown at a normal angle so that viewers will find parallels to the real world in the use of perspective. During the visualization process, you can arrange elements of a virtual model in combination with virtual light sources from an advantageous perspective. These form the render scene.

\\Hint:
Although sunlight should generally be understood as point light, it has an effect similar to parallel light when it strikes the Earth's surface due to the sun's enormous size.

\\Tip:
Some CAD programs include sun and shadow studies as a design aid (e. g. Graphisoft ArchiCAD and Nemetschek Allplan). To set the position of the sun, you enter into a dialogue box the desired latitude, longitude, time of day and time of the year. This allows you to simulate the lighting conditions in any place on Earth at any time and thus provides you with valuable information on the effects of building orientation.

Fig.85:
The light source produces a hard shadow which, while not geometrically precise, requires little computation time.

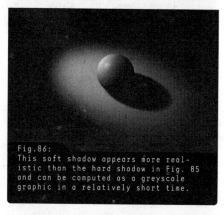

Fig.86:
This soft shadow appears more realistic than the hard shadow in Fig. 85 and can be computed as a greyscale graphic in a relatively short time.

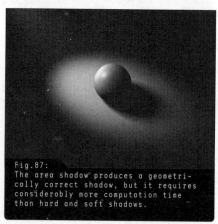

Fig.87:
The area shadow produces a geometrically correct shadow, but it requires considerably more computation time than hard and soft shadows.

The use of scenes often saves time because they can be stored as a template and thus do not have to be recreated for every visualization computation. As a rule, the render scene is visualized using a virtual camera, which can be used and adjusted within the application parameters like a real camera and defines <u>eyepoint</u>, <u>aiming point</u>, <u>focal distance</u> and <u>camera angle</u>. › Fig. 88 The eyepoint corresponds to the "eye level" of the observer in the virtual space. Like the aiming point of the view, it can be defined via coordinates in three dimensions. › Chapter Coordinate systems

```
\\Hint:
When using rendering software, you can gener-
ally store scenes and retrieve them at a later
point. A scene is an arrangement of virtual
objects and light sources shown from a selected
perspective that can normally be adjusted by
means of virtual cameras (see Chapter Perspec-
tive and virtual camera).
```

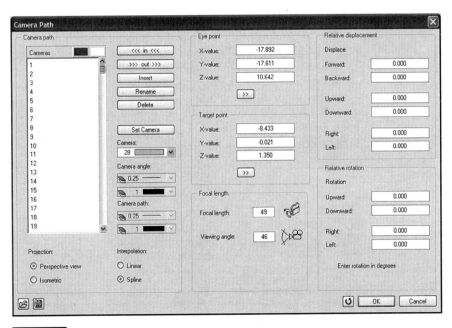

Fig. 88:
Setting options for the parameters of a virtual camera (Nemetschek Allplan)

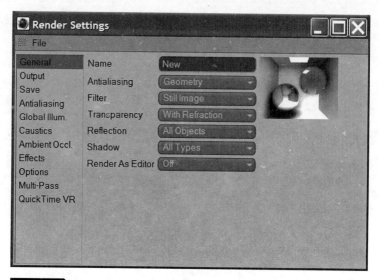

Fig. 89:
Example of a dialogue box for rendering presettings (Cinema 4D)

The camera angle and the focal distance are geometrically dependent on each other and determine the visible field of the camera perspective: reducing the focal distance allows you to extend the visible field even in cramped spatial situations. However, to avoid the "fisheye" effect, you should not reduce the focal distance too much. If possible, eyepoint and aiming point should be located at the same height so that vertical edges can be depicted parallel to one another.

\\Hint:
Many rendering programs offer a variety of rendering engines (rendering types) that differ in terms of their basic computational methods. They include Raytracing, Phong Shading, Gouraud Shading, OpenGL and Z-Buffer. A simple way to compare image quality, calculation times and the advantages/disadvantages of each engine is to generate renderings of the same object.

RENDERING PARAMETERS

The quality of a visualization is based not only on an optimum arrangement of objects, surfaces and light sources, but also on the technical quality of the image. This quality depends on a wide variety of parameters, which must be properly harmonized. One method of defining parameters is by the presettings of the rendering software. We provide a basic explanation of these parameters below. › Fig. 89

Generally speaking, the term "rendering" refers to the generation of new data from suitable "raw" or source data. In the specialized field of computer visualization, it describes the process of converting a vector graphic into a bit-mapped graphic.

Resolution

The resolution of a pixel graphic is based on the number of pixels in a given area and determines the quality of the image. Normally, resolution is expressed by the abbreviation dpi (dots per inch) as the width x height of the image. The more pixels there are in an image, the finer the resolution. Conversely, an image with fewer pixels looks grainier and the depiction of finer details is not as crisp. › Fig. 90

In conjunction with the total number of pixels, the resolution determines the effective size of a pixel graphic as expressed in centimetres. At a resolution of 300 dpi, a 1000 × 1000 pixel rendering has a square shape measuring 8.47 cm × 8.47 cm. If the resolution is reduced to 150 dpi, the size of the image increases significantly to 16.94 cm × 16.94 cm, and the same number of pixels is distributed over an area that is four times as large. This is why, in contrast to a vector graphic, a pixel graphic cannot be enlarged without a loss of quality, since only the size of the individual dots increases. This affects the quality of the image, which becomes "pixelated" and has a coarser resolution.

\\Hint:
A vector graphic is a two- or three-dimensional computer image composed of different drawn elements, each of which is defined by coordinates. In contrast to bit-mapped graphics (see below), vector graphics can be enlarged to any desired size without a loss of quality. Compared to other graphics formats, they also require little memory volume.

\\Hint:
Bit-mapped graphics (also called pixel graphics) consist of an arrangement of individual pixels. A pixel is the smallest unit of a pixel graphic and contains a particular colour value as graphic information. Put differently, a pixel graphic is a combination of many pixels that is comparable to a mosaic of many small elements.

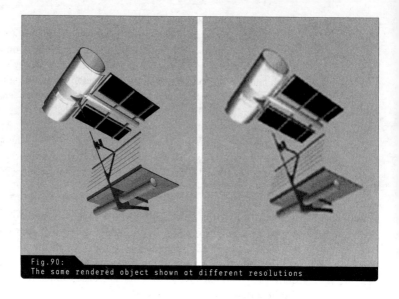

Fig.90:
The same rendered object shown at different resolutions

You need only a little experience to determine the required computational times based on the planned resolution of a rendering. If a larger number of pixels need to be calculated for a computer graphic, the calculation will require more time and effort. Renderings that represent intermediary stages of the design generally require a lower resolution than final presentations and their rendering times will be shorter too.

Anti-aliasing

When converting a vector graphic to a pixel graphic, you can use anti-aliasing to avoid a jagged depiction of diagonal lines (the so-called stair effect). A pixel graphic is only capable of portraying horizontal and vertical lines without great difficulty. If a line is slanted, "stairs" appear because slanted lines are portrayed by offset square pixels in a pixel graphic. › Fig. 91

The same effect can be seen in all kinds of round shapes and even in typefaces. The grainier the resolution, the larger will be the individual pixels – and the more pronounced the stair effect. Anti-aliasing interpolates and partially weakens the colour values of the individual "steps" so that they are not as distinct. The level of anti-aliasing has a substantial effect on the time needed to compute a rendering: if the round shapes and diagonal lines need to be portrayed crisply, the calculation will take longer. This is why maximum anti-aliasing should only be used for final presentations.

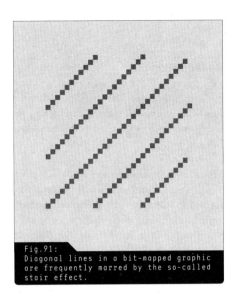

Fig.91:
Diagonal lines in a bit-mapped graphic are frequently marred by the so-called stair effect.

Aside from the wealth of geometric information required for a virtual model, the rendering settings, in particular, exert a great influence on the time needed to compute a visualization. For this reason you should, if possible, consider the purpose and the intended quality of a visualization or rendering result before you begin the rendering process. > **Appendix, Table 2**

\\Tip:
You can also increase resolution to diminish the stair effect. In this case, the diagonal lines and edges are calculated using an especially large number of pixels and there is a finer depiction of the inevitable jagged patterns.

\\Tip:
You should first test the rendering settings on a detail from the virtual scene in order to examine, in advance, the effects of light and materials. Complex models, in particular, may require a rendering time of several hours.

DATA FLOW

PROGRAM LIBRARIES

CAD programs normally feature integrated program libraries with an array of symbols, document templates and predefined building components. Architectural CAD programs include sundry objects that can serve as standard drawing elements or that can be changed to reflect your own style and ideas. › Fig. 92

You can also create templates from objects you have drawn yourself and use them in more than one project. It will save you a great deal of time to use objects defined in this way since no intensive research is required; and you can integrate objects into drawings, and copy and edit them as you require, instead of drawing them from scratch in the future.

Symbols

There are symbols to represent both interior furnishings (furniture, bathroom fittings, kitchen elements etc.) and elements of landscape design (trees, plants, cars, people etc.). Additional icons depict objects in the field of structural engineering, building services technology and the planning disciplines linked to architecture.

Building components

The use of standardized components saves a great deal of work, particularly when you create working drawings, because you can then select drawing elements like steel components (beams, pipes etc.) and connectors (screws, bolts etc.) from a catalogue, incorporate them directly into the drawing and edit them there. Even complex components such as stairways, windows and doors can be stored and used in this way. › Chapter Architectural elements

Pixel images

In addition to elements in vector graphics, pixel images can also be incorporated into libraries and administered there. These images may

> \\Tip:
> Using a scanner, you can scan both freehand sketches and other hand-drawn technical drawings, and then import them as pixel graphics into CAD drawings. Conversely, you may also print out mid-project CAD drawings, modify them by hand using transparent paper, and then transpose the results to CAD. This is one way to combine the advantages of CAD and hand drawings.

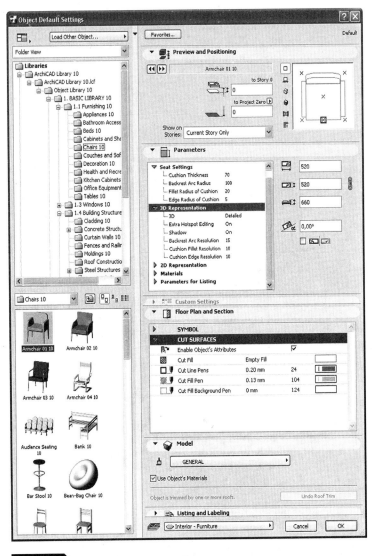

Fig.92:
Example of a program library (Graphisoft ArchiCAD)

include digital photographs, scans and renderings that are systematically imported, stored, and – if needed – integrated into a drawing.

Document templates — Drawing frames can also be imported and edited, and the same applies to standardized lettering and drawing headers in the respective scale. This means you need to create customized document and printing templates only once. These are then stored for later use in the program library.
> Chapter Printing and plotting

CAD INTERFACES

CAD interfaces facilitate the exchange of data between different programs. Although all CAD systems generally work with vector graphic information, most programs employ their own file formats, which are often incompatible and impede data exchange between different CAD programs. DXF has largely caught on as the standard data exchange format for architectural drawings, as it can be read and written by nearly all CAD software and even by a few graphics programs.

DXF is short for <u>drawing interchange format</u> (or drawing exchange format) and was developed by the company Autodesk to exchange CAD data.

If a DXF file is imported into a CAD program, different options can be chosen to adapt specific properties of the vector graphic data. These include unit, scale and the two- or three-dimensional mode of transmission. > Fig. 93

A number of architectural CAD programs not only use DXF and their own data formats, but also provide interfaces to select programs in order to ensure trouble-free data exchanges. This is particularly true in the field of rendering programs – the additional software necessary to guarantee outstanding visualization results. For instance, Nemetschek Allplan uses a special data format to export three-dimensional models to Cinema 4D rendering software, Autodesk Architectural Desktop offers a comparable

---

\\Hint:
Exchanging different CAD data is not problematic and rarely goes off perfectly. Important details specific to a CAD system are often lost or may not be displayed in the same way as the original exported information. This can become glaringly evident in the case of typefaces, program-specific symbols and dimensioning data, since they rarely have an equivalent in the target system.

\\Hint:
Special programs that read and write data in several file formats and thus facilitate imports and exports between different CAD programs provide an alternative to DXF-based data exchanges. That said, you may have difficulty transmitting data if you lack the necessary experience.

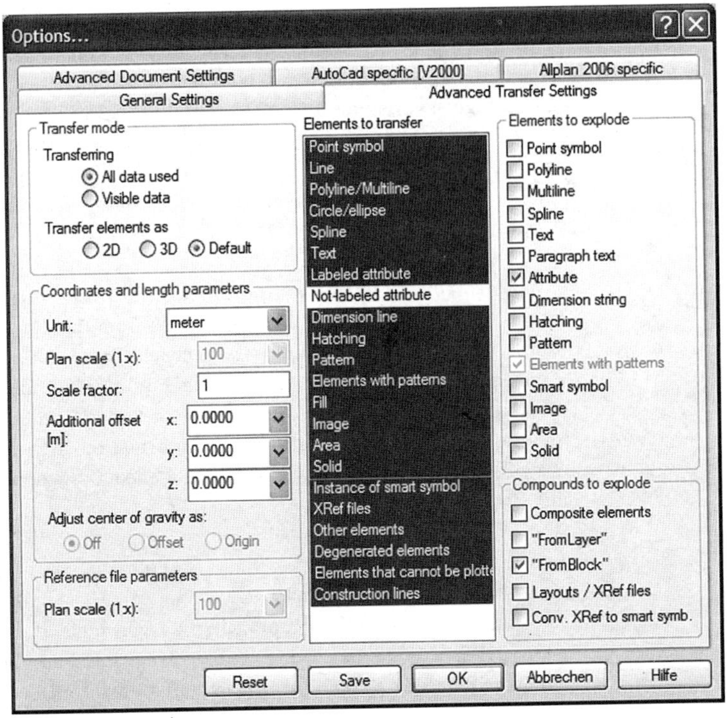

Fig. 93:
Example of a dialogue box for exporting and importing DXF files (Nemetschek Allplan)

interface to 3D STUDIO MAX/VIZ, and Graphisoft ArchiCAD has developed an interface for Artlantis.

TAI

TAI, short for tendering, awarding and invoicing, represents an important area of building construction. After architects create the design and construction documentation for an architectural object, bids or tenders are invited for the individual construction jobs – in other words, this work is described in written form, its scope is specified, and this information is made available to construction companies together with relevant drawings so they can submit a tender. After a quotation is received, the construction company is selected based on the specified prices and quality standards, and the construction work is commissioned. Finally, after the construction work is completed, accounts must be settled for the costs incurred.

Special tendering, awarding and invoicing programs make it easier to manage quantities and costs in building construction. When architects create a structure as a complete virtual model using CAD, they can categorize the quantities of the individual building components according to materials and skilled trades. > Chapter Architectural elements This information can be imported directly to the TAI programs and edited there, eliminating the need for time-consuming calculations of quantities by hand.

From the CAD system to TAI software to the construction site and back again

Ideally, data will flow from the CAD drawing to the tendering program. Setting quantities links components from the CAD drawing directly to the TAI program, allowing users to check the allocation, quantities and dimensions of components and items in both the CAD and the TAI system. This allows you to compile a list of contract specifications efficiently and monitor their completion effectively. Information from the CAD program (attributes, quantities etc.) can be calculated and organized in corresponding lists.

PRINTING AND PLOTTING

After a drawing is completed in design mode, it can be printed or plotted. While large-size CAD drawings are printed on plotters, small formats are printed on standard DIN A4 and A3 printers.

Drawings and plans are usually prepared for printing in special windows that allow you to arrange different drawing content and pixel images on a page, as well as subsequently labelling them and making other settings relevant to printing.

Print scale

Here you make the final specifications regarding the printing scale and the required paper size. You also have an additional opportunity to examine, prior to printing, the effect of scale-related settings such as line

\\Hint:
Plotters are large-format printers, usually DIN A1 or A0, on which plans are printed on roll paper. Roll widths normally measure 61.5 cm for A1 rolls and 91.5 cm for A0 rolls. Plan sizes can be entered individually into the CAD programs so that every plan format can be used inside the given roll width.

\\Tip:
When printing on a plotter, you can usually choose from three different quality settings. A low quality level will print much more quickly than a higher setting, but the print will have a lower resolution and not be as sharp. Lower settings also use less ink.

width and font size. In this context it is important to make sure that the reference scale matches the printing scale, since characters and other scale-dependent features of a drawing might otherwise be too small or too large. › Chapter The virtual drawing board

For instance, if a plan is supposed to show a façade section together with corresponding detail views, it is possible to present drawings in different scales on a single architectural plan. In this case, each individual drawing is depicted in the required scale.

Paper formats

Plan formats with different aspect ratios are available for presentation purposes. It usually makes sense to choose a common paper format for architectural drawings (e.g. the ISO/DIN A series), since it is easy to produce them. Large formats are needed for construction drawings, while for case of detail drawings, it is generally better to use a format that can be copied on most copy machines, such as DIN A3 or A4.

A drawing frame created for the selected paper format generally defines the area taken up by the drawing, and shows the edge that can be used to subsequently cut the drawing. The drawing header contains technical information such as a precise description of the content (floor plan, section, view etc.), the scale (M 1:50, 1:100, 1:100 etc.), as well as printing date, author and other supplementary data. › Fig. 94

The drawing frame and header can be created as templates in the required scale, making it possible to use the basic settings for other projects as well. › Chapter Program libraries

Virtual print

You can also print drawings and plans "virtually" in the form of files. In this case, you do not transmit the vector graphic information in a drawing to a real printer or plotter but store it as a plot file using the printer software. A plot file contains all the drawing information that is required

\\Hint:
There are sometimes significant differences in the plan design options provided by CAD programs, and their ease of use can differ too. A few providers offer supplementary modules specially developed for plan design and printout (e.g. Nemetschek Plandesign).

\\Tip:
This method can be employed in a number of situations – for example, if a student does not own a plotter and needs to make a print on the university's plotter or at a professional copy shop. Data can be printed on the plotter without installing the corresponding CAD program on the attached computer.

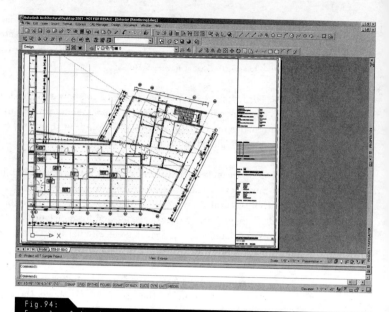

Fig. 94:
Example of drawing window (Autodesk Autocad)

by a specific printer or plotter to print out the drawing at a subsequent point in time without using the CAD software's printing environment.

PDF

PDF stands for "portable document format" and is a common data exchange format developed by Adobe Systems that can contain both vector graphic and bit-mapped information. Many CAD systems allow you to create PDF files as an export function. Alternatively, you can install a virtual printer (e.g. Adobe Acrobat Distiller) in much the same way as a real printer and then select it for printing. It will create a PDF document containing all the graphic information of the virtually printed drawing.

Pixel graphics

A few CAD systems allow you to export drawing content as pixel graphics. These can be stored both in different file formats and, if required, in user-defined resolutions, two examples being the JPEG or/and TIFF formats. A distinguishing feature of the JPEG format is that the bit-mapped graphic information is mostly stored in compressed form so the file sizes are accordingly small and easy to manage. The TIFF format requires much more memory, but it contains a great deal more graphic information and is better suited for additional editing steps (e.g. image-editing programs).

SYSTEM REQUIREMENTS

HARDWARE

Even though, as the user of a computer, you do not necessarily need to know much about its inner workings, we are including the following sections to provide some basic information about a number of hardware components.

Central processing unit The central processing unit, or CPU, is the heart of the computer and controls all its other components.

Mainboard The mainboard (also known as "motherboard") is the central circuit board of the computer and has attachment points for its processor and memory modules as well as for supplementary cards such as graphics, sound and network cards. These components can also be integrated directly into the mainboard, in which case they are said to be "onboard."

RAM – working memory RAM, short for random access memory, is your computer's working memory.

Graphics card The graphics card in a PC controls the screen display, ensuring, among other things, that data are computed quickly for visualizations. For CAD systems, it is advisable not to have an onboard graphics card since the relatively slow RAM will be used for graphics processing and the other processes may be slowed down.

Hard drive A hard drive is a storage medium that writes data on the magnetic surface of a rotating disc. The surface is magnetized based on the information being recorded. Even if a hard drive has sufficient memory capacity for all CAD data and is a relatively secure storage device, you should regularly secure CAD data on external media such as CDs and DVDs to protect against possible data loss.

Monitor The screen size of a monitor is normally indicated in inches and is based on the screen diagonal. The screen should not be too small since it is used to display the virtual workspace. Alternatively, you can use two screens: if you display all the interface's control functions on one, you can use the virtual drawing area on the other without restriction. Please keep in mind, though, that your graphics card must support this setup.

Special input devices Aside from the keyboard and the mouse, other input devices have been specially designed for the virtual workspace. The best-known are the joysticks used to play three-dimensional computer games.

SpaceMouse   The SpaceMouse functions like a joystick and can efficiently control virtual views. Additional keys can be programmed to execute drawing and tool commands.

Sketchpad   The sketchpad, also called a "digitizing board," digitizes data entered with a pen-like pointing device. Since the sketchpad provides a much higher resolution than a standard computer mouse, you can use it to draw an object in much the same way as you would draw on a piece of paper with a pencil. When assigned the appropriate scale, the drawing area of the user interface will correspond to that of the sketchpad.

Touchscreen   New touchscreens can also be used for drawing, like sketchpads. The touchscreen is not set up vertically like a monitor in front of the user, but is used as a horizontal work surface to create and display drawings.

SOFTWARE

If at all possible, you should choose CAD software to match existing hardware. The use of modern software on old computing systems can be problematic, since these may not provide the computational capacity required to execute the software operations adequately. You must also have a suitable <u>operating system</u>, which controls the computer's fundamental processes. This is a basic requirement for running all application software. The most popular operating systems are Windows (Microsoft), Mac OS (Apple) and, increasingly, LINUX, which you can largely use without a license. Not all operating systems are suitable for all CAD systems, and you should pay close attention to the software providers' specifications.

There is a large selection of CAD software on the market, and the various CAD programs often function differently and have significant price differences. From freeware and shareware to costly specialized programs, the offerings are immense. › Appendix, Table 3

An alternative to the expensive complete versions of CAD software are the demo and student versions that a few providers make available to consumers. The student versions are inexpensive compared to the normal versions, and you can even use the demo versions for free for a limited period. Even so, it is prohibited to use them professionally. A demo version may make sense to take the first few steps with CAD since it allows you to try out the software and evaluate its strengths and weaknesses.

Selecting software   Selecting a CAD system depends on a number of factors: the (architecturally relevant) functions, the special properties of operation and use, the cost and the required hardware. An additional criterion is the CAD user's personal or professional environment since data exchange with project partners – to name just one example – should be as trouble-free as possible and is best ensured by exchanging data from one and the same software program.

Program-specific literature is available for the various CAD programs in addition to training courses at universities, adult education centres and private institutions. If you do not want to learn CAD on your own – which is, of course, entirely possible – you can take a course that suits your particular level of expertise.

It is advantageous to learn not just a single CAD system, but to master at least the basic principles of other programs so that you are not restricted to particular system requirements and remain flexible in your work. Although the operation of many CAD programs differs, they all have the same underlying foundations. Once you have a good command of these basics, you can apply this knowledge to other programs and learn them more easily.

\\Hint:
Different software providers will make recommendations concerning hardware configurations. These should be regarded as minimum requirements when you purchase your CAD system. The software may function on the recommended hardware, but it may not be able to carry out work processes involving large volumes of data. This is why it is worthwhile to get advice from a specialized vendor on which computer system to buy.

\\Tip:
Freeware and shareware can be downloaded from the Internet but in most cases only offer limited basic functionality.

DESIGNING IN DIALOGUE WITH THE COMPUTER

The computer in general and drawing with CAD in particular streamline work processes and provide a basis for efficient and exact approaches to creating and managing digital drawings and other data. Associative components and virtual models not only simulate reality, but also influence it directly by offering a new manner of designing and systematized planning process. But even with all its benefits, this working method has a few drawbacks. Designing with CAD usually entails drawing with a mouse, which cannot be guided as precisely as a pencil. Also, you only have indirect contact with the drawn object via the entry and output devices—an aspect of the work that should never be underestimated. When drawing with CAD, beginners, in particular, are confronted with many operations that may initially seem complicated and can influence their mental processes, particularly when they engage in creative design.

By contrast, drawing by hand is often an intuitive process in which personal ideas are implemented directly and sometimes even largely unconsciously. The results on paper are continually influenced while you draw, and the representational method, which is, not least, dependent on experience and feel, is aligned with the drawing material and the represented object. This is another reason that you should incorporate hand drawings into the design process and, if possible, enhance the use of the CAD application with hand sketches.

Outlook

CAD has become a firm feature of the day-to-day work of designers, and it will continue to grow in importance in the future. In the field of architecture, the developments playing an especially important role are those that allow all project participants to have direct access to an object as a virtual model – from the creation of the design and working drawings to the construction of the object. This makes it possible to evaluate, print and even modify the data and drawings that are relevant to planning and construction directly. It lays a foundation for virtual planning that is always up to date and, if need be, reflects the architectural object down to the finest detail. In addition, even when it accompanies an ongoing development process, CAD influences the development of new approaches to planning and allows for the design of architectural structures that could not completely be depicted by hand drawings, including amorphous and other free forms. In this context, complex structural calculations or the computation of complicated component geometries are an important basis for the design and construction process. CAD represents much more than an easy-to-use drawing tool. It is both a comprehensive instrument and an important building block in the development and future of architecture.

APPENDIX

CHECKLISTS AND OVERVIEW OF SOFTWARE

Table 1: Sample layer structure for designing a building

Environment	Building
	Development
	Trees
	Piece of property
Design	Design grid
	Layouts
	Views
	Sections
Loadbearing structure	Construction grid
	Foundations
	Exterior walls
	Interior walls
	Columns
	Beams
	Stairs
	Ceilings
	Roof structure
Building finishing	Grid
	Lightweight construction
	Electrical installation
	Heating and plumbing
Furnishings	Bathroom fixtures
	Furniture
	Objects
	Flooring

Drawing information	Dimensioning
	Labelling
	Areas and rooms
	Room label
	Hatches
	Patterns
	Fillings
	Symbols
	Markers
	Fixed points
Plan layout	Frame
	Illustrations
	Scans
	Renderings

Table 2: Checklist for a computer visualization

Capacity	Hardware performance
	Software options
	Timeframe provided
Purpose of visualization	Design tool
	Intermediary presentation
	Final presentation
Level of detail in visualization	Close-up
	Building portrait
	Extended context (e.g. aerial shot)
Rendered scene	Choice of perspective, camera settings
	Image format
	Rendered object

	Lighting scenario
	Background or surroundings
Rendering parameters	if needed, rendering type (e.g. Raytrace, Phong, Gouraud, OpenGL, Z-Buffer etc.)
	Resolution
	Anti-aliasing
Surfaces	Colours
	Textures
	Reflexions
	Transparency

Table 3: Overview of CAD software

Program	Homepage
Allplan	www.nemetschek.com
ArchiCAD	www.graphisoft.com
ArCon	www.arcon-software.com
AutoCAD/Architectural Desktop/ Inventor/Revit Building	www.autodesk.com
BricsCad IntelliCAD	www.bricscad.com
CAD	www.malz-kassner.com
CATIA	www.catia.com
ideCAD	www.idecad.com
MicroStation	www.bentley.com
Reico CADDER	www.reico.de
RIB ARRIBA® CA3D	www.rib-software.com
SketchUp Pro	www.sketchup.com
Spirit	www.softtech.com
Solid Edge	www.ugs.com
TurboCAD	www.imsi.com

Table 4: Overview of rendering software

Program	Homepage
3ds Max/VIZ	www.autodesk.com
Artlantis	www.graphisoft.com
Cinema 4D	www.maxon.net
Maya	www.alias.com
mental ray	www.mentalimages.com

PICTURE CREDITS

Figures 1, 4a, 12, 65, 88, 93:	Nemetschek AG, Munich – Allplan 2006
Figures 2, 3, 4b, 10, 11, 24, 69, 71, 92:	Graphisoft – ArchiCAD 10
Figures 4c, 9, 25, 68, 94:	Autodesk GmbH Deutschland – Architectural Desktop 2007
Figures 53–55:	Bert Bielefeld, Isabella Skiba
Figures 73, 74, 76, 77, 78:	ch-quadrat architekten
Figure 75:	Frank Münstermann
Figure 81:	HKplus architekten
Figures 82–87:	HKplus architekten / the author
Figure 89:	Maxon Computer GmbH – Cinema 4D R10
All other figures:	The author

CAD：定义及应用领域

　　CAD 是"计算机辅助设计"（Computer-Aided Design）的英文首字母缩写，即在计算机的帮助下进行设计、绘图。我们可以选择种种 CAD 程序（软件）来绘制二维或三维的图形，所需要的绘图工具仅仅是电脑的键盘和鼠标等一些电脑输入设备。绘制的图形通过诸如电脑显视器或打印机等设备进行显示。CAD 这种以 IT 技术为根本的应用在包括工程设计的很多领域中起到了非常重要的作用。本书集中介绍了 CAD 在汽车、设备工程、结构工程及建筑学领域的应用。除了工程制图，CAD 工具也被应用于创建各种各样的虚拟模型。在 CAD 工具的帮助下，人们可以虚拟建筑物在建成之后的效果，即便是虚拟气候变化和光影效果也不在话下。有些 CAD 工具亦可以进行荷载模拟或流体模拟，这可是给很多工作帮了大忙的。

　　第一个 CAD 工具被成功开发于 20 世纪 60 年代，主要用于建造飞机。20 世纪 80 年代以来，随着个人计算机的普及以及电脑工作站成本的降低，越来越多的用户开始能够应用这些 CAD 工具了。自 21 世纪初发展起来的性能强大、制式统一的计算机系统大力促进了 CAD 工具的发展，这些工具更加有效，成本更加低廉，并且能够适应各种不同的需求。

　　尽管几乎所有的 CAD 工具都有差不多相同的原理基础，但是在操作和应用上还是存在一些区别的。现在市场上有很多关于 CAD 的著作，但基本上都是针对特定的 CAD 工具的。而本书的出发点与上述著作不尽相同，本书的主要目的是提供一个通向 CAD 应用的通道，从基础知识、应用定位等方面入手，为读者如何选择一个符合自己使用目的的 CAD 工具提供帮助。

提示：
　　为了能够实际地与现存 CAD 系统联系起来，本书选择了一些建筑 CAD（CAAD）的典型功能作为实例。具体名称请见本书附表。

虚拟绘图板

当你使用CAD工具绘图时,电脑的显示器便是你的图板,鼠标和键盘取代了你的铅笔。在各种绘图功能的支持下,通过鼠标键盘绘制图形变得十分简单(见"绘图元素")。

显示比例

在使用CAD绘图时,我们通常采用1:1的绘图比例进行绘图,也就是说,按照被绘制物体的实际尺寸进行绘图。比如,对于一个10m长的墙,我们用CAD绘制它时采用的长度也是10m。为了在电脑屏幕上显示这堵墙,我们必须选择小比例尺以适应电脑屏幕的大小,此时,显示比例这一概念便应运而生——显示比例描述了绘制对象实际尺寸与在电脑屏幕上实际显示尺寸之间的比例关系,这一显示比例是随着屏幕显示大小的变化而改变的。通常意义上所说的绘图比例只有在CAD出图的时候才被涉及(例如打印),但是,绘图比例这个概念还是绝不能被忽视的。

参考比例

在CAD绘图中,参考比例是用来描述打印尺寸与被绘制物体实际尺寸之间的比例的(例如1:20、1:100或者1:500)。由于诸如文字一类的绘图元素是脱离于被绘制物体而相对独立存在的,为了能够保证在打印绘图中清楚地显示这部分绘图元素,应用于这部分绘图元素的绘图比例需要参照打印比例酌情而定。也就是说,在绘图时,需要通过CAD工具软件定义字体与其他绘图元素之间的比例(见"打印和制图")。

度量单位

在进行工程制图时,建筑师采用公制度量单位以便每个人都可以理解认知。在中欧,毫米、厘米和米是公制度量单位。有些英语国家的人们还是喜欢沿用英制度量单位的,但是出于对国际工程标准化的考虑,这些国家也开始越来越多地使用更加容易换算的公制度量单位。不过不管怎样,使用什么度量单位绘图那是你自己的事情,至少CAD工具不会给你制造什么障碍。毫米、厘米、米、英寸、英尺等等都可以作为最小度量单位出现。

用户界面

CAD的用户界面由几个部分构成,以下会对几个部分详加解释。

各种CAD绘图工具在CAD主程序界面上以按钮符号和菜单的形式显示(图1)。我们可以应用这些绘图工具很简单地在CAD界面上创建图形。

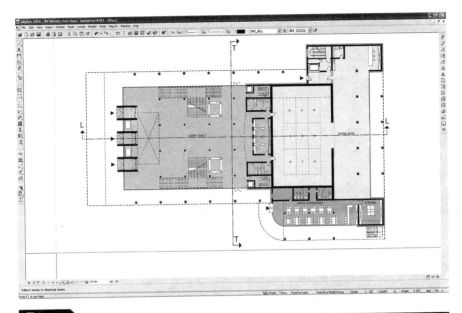

图 1：
CAD 用户界面（Nemetschek Allplan）

绘图区域　　　绘图区域是用户界面中最重要的一个区域，我们可以在这个区域任意绘制、修改二维或三维绘图。简而言之，绘图区域也就是我们手绘图时的那张绘图纸，有所不同的是，CAD 的绘图区域是一个虚拟的工作区，给我们提供了更大应用空间和多样的虚拟工具。

选择工具和　　鼠标是最常用的指点定位设备，鼠标控制光标在用户界面移动，
绘图工具　　　配合用户的指点动作完成选择和绘图行为。光标在电脑屏幕上的表现为一个箭头或十字交叉线，抑或其他 CAD 定义的形状。这里所说的光标形状会根据不同的功能切换（例如选择或绘图等）而随之变化，我们有时可以根据光标的形状来判断当前的功能状态。

注解：

当我们想要让屏幕显示尺寸与实际尺寸相同时，我们必须将显示比例与参考比例设为相同。有些 CAD 程序将显示比例与参考比例以百分数的形式联系起来，当显示比例是参考比例的 100% 时，显示真实尺寸。我们可以在设计过程中打印例图，以便对比例效果进行实时把握。

提示：

除了鼠标、键盘以外，我们还可以选择手写板和空间鼠标等作为输入设备。但是这类设备在建筑 CAD 领域的应用并不太多（见"系统需求"之"硬件设备"）。

电子笔

我们可以自行定义 CAD 里面的电子笔，使其具有不同的宽度、线型及颜色（图2）。可以在开始绘图时定义这些项目，亦可在绘图进行中修改这些项目。这会使整个绘图过程变得非常清晰，绘图者可以使用不同的绘图元素来定义不同的绘图单元以便进行区分。由于 CAD 已经事先将不同的线宽与不同的颜色挂钩，我们可以根据线型特征很清楚地分辨不同的绘图单元。这样一来，在绘图时便可以对各种绘图单元事先定义，从而为日后的打印出图埋好伏笔（见"打印和制图"）。由电子笔绘出的线型会一五一十地表现在打印输出的图纸上。

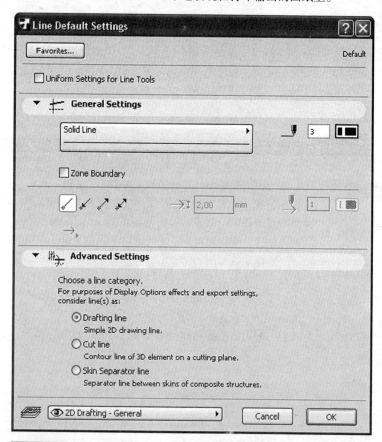

图2：
我们可以对电子笔所绘的线宽、线型及颜色等属性进行定义（Graphisoft ArchiCAD）

另外，许多 CAD 程序为我们提供了显示真实比例线宽的功能。开启这一功能，我们就会在电脑屏幕上看到与打印输出一模一样的线宽，更加直觉一些。在进行视觉效果处理时，电子笔也可以起到一定的作用（见"视觉效果"）。

菜单栏

菜单栏位于绘图区域上方。例如，在 Windows 系统中，我们可以使用菜单栏里的"文件"（File）栏目来创建、保存绘图文件。我们可以使用鼠标、键盘激活菜单栏中的各种功能，以及配置 CAD 程序等。

工具栏

工具栏中有很多按钮，各个按钮都对应着不同的绘图工具（图3）。我们将鼠标停放在按钮上一会儿，按钮所对应的绘图工具名称便会显示出来。我们在使用 CAD 时，可以对工具栏里面的工具进行配置，根据自己的需求显示或隐藏不同的绘图工具按钮。

关联菜单

关联菜单所列项目与用户正在应用的绘图功能有所关联。在默认模式下，用户点击鼠标右键就会调出关联菜单。关联菜单使用户很快捷地完成重复、取消某一命令或激活一些绘图工具等动作。

对话框

对话框实际上是用户与 CAD 进行信息交互的一个窗口，用户可以了解到自己选定的一些功能指令的具体执行信息，也可以通过键盘在对话框中输入绘图命令或者各种绘图功能所需要定义的数值。对话框是伴随着各种绘图功能存在的，在用户打开 CAD 界面的时候就能看到（图4）。

P15

坐标系

CAD 软件的整个图形创建平面是一个以坐标系为基准的虚拟绘图平面，整个平面可以被看作为一个由水平和竖直线交织分格的方格纸。图形创建平面也可以被看作是由坐标点构成的平面，在绘制图形时，这些坐标点可以帮助定义各种绘图元素的位置以及形状（见"绘图元素"）。

我们通常用到的<u>直角坐标系</u>为图形创建平面的基础坐标系，这个由 X 轴和 Y 轴构成的二维坐标系可以描述任何点到原点的距离（图5）。

注解：
工具栏可以粘附在绘图区域边缘，也可以拖放到绘图区域里。尽量扩大绘图区域是比较明智的做法，然后待熟练使用工具功能之后，合理地安排工具栏位置，以便快捷地找到想要的工具按钮。

提示：
直角坐标系并不是惟一选择，极坐标也可以应用在 CAD 系统中。一个点在极坐标中可以表示为该点到原点的半径及与固定轴的夹角。

图3：
工具栏示例（Graphisoft ArchiCAD）

图4：
CAD对话框（Nemetschek Allplan，Graphisoft ArchiCAD，Autodesk Architectural Desktop）

从点引出的分别与X轴和Y轴平行的虚线可以帮助确定该点的位置坐标。

更进一步讲，我们可以通过确定一条线段的两个端点的位置坐标来确定该线段的几何位置。这就是为什么在直角坐标系绘制线段时必须要指定端点位置的原因（图6）。

提示：
通常情况下，图5、图6所示的坐标系在CAD的绘图区域是看不到的。有些CAD会在屏幕边角示意性地显示坐标系标志，指示坐标轴方向。

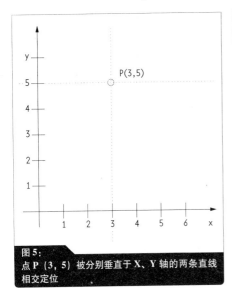

图 5:
点 P (3, 5) 被分别垂直于 X、Y 轴的两条直线相交定位

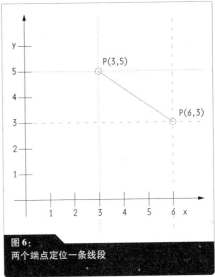

图 6:
两个端点定位一条线段

绝对坐标   绝对坐标是以上文提到的 X 轴和 Y 轴相交的原点为基准的,当我们要定义一个点的确切的位置时需要用到这一坐标。

用户坐标系   当绘制一张很大的图纸时,仅仅使用绝对坐标会给计算和输入带来很大的麻烦。

我们会发现,有很多的 CAD 程序会允许用户在绘制图形时任意地移动或重新定义坐标系原点,这其实就是从简化工作这一角度出发的。用户甚至可以将整个坐标系旋转任意一个角度,这使得用户在绘制不垂直于原始坐标轴的直线时更加方便。以下我们把这种可以由用户自行定义的坐标系称为用户坐标系(UCS)。

相对坐标   一种灵活交替运用坐标系的方法催生了相对坐标的形成,我们可以将刚刚定位好的点作为下一个点的参照原点,这便是相对坐标的精髓所在(图7)。

P17   **透明的平面——分层原则**

CAD 绘图中通常会有很大的信息量,当图纸很大且复杂程度很高时,你会发觉工作难度非常大。很多不同的绘图单元彼此相邻并互相联系时便会出现这种状况,CAD 通过使用类似透明图纸的透明图层依次层叠来解决此类问题(图8)。这种所谓的图层在不同的 CAD 程序中有不同的叫法,例如在 Nemetschek Allplan 程序中,图层被称为"图卷"。

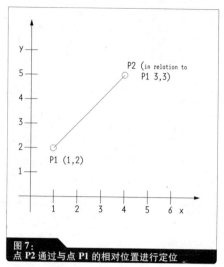

图7:
点 P2 通过与点 P1 的相对位置进行定位

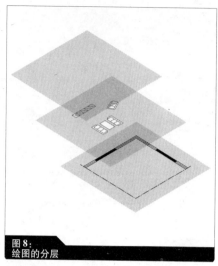

图8:
绘图的分层

　　另外，层结构对绘图结构的组织和编制起到了很重要的作用，用户只需要通过一个对话框，就可以创建任意数量的图层，并且可以方便地通过图层管理工具管理各个图层上的内容属性。比如说，一个建筑会有很多不同的组成元素，这些组成元素反映在建筑图上时，如此简化工作十分行之有效：将建筑的内外墙、标注、文本、剖面线等等都分别绘制在不同的图层上，这使得对这些元素的编辑修改或分类检索都十分方便清晰（图9及附表1）。

实例：
　　将隔墙绘制在平面布置图中，内部装饰绘制在其他图层中，两种绘图元素都可以分别在各自图层中显示、编辑。必要时，两个图层中的元素还可以进行合并。

注解：
　　一个设计项目开始时，非常有必要将所有图层的安排方式整理清楚。比如说，要将所有图层分门别类地事先定义好，即便大部分图层都要等到后期才会用到也必须这么做。另外，如果一个绘图对象的组成成分分别存在于几个图层，可以将几个图层组成一个图层群。按照这种方法设置图层，会有事半功倍的效果（图10）。

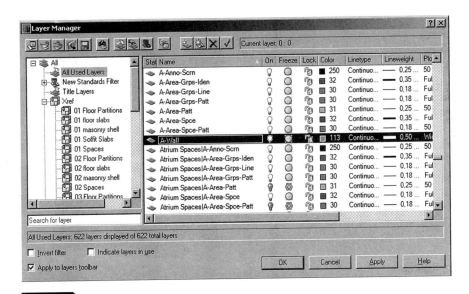

图9：
CAD 图层管理界面（Autodesk Architectural Desktop）

过滤功能

　　另外，我们可以任意显示或隐藏一个或多个图层中的内容，也可以将一个图层锁定，保证该图层中的内容不被更改。

　　上文所述的图层的分层管理功能不仅使绘图工作更加容易操作，而且还能够快速有效地检索绘图中的各种绘图元素。而这种检索只是以一个图层的属性进行检索，在某些情况下，一个图层内的绘图元素的属性并不一定完全一致，CAD 对此提供了一个更加完善的检索功能，我们称之为"过滤功能"，使用过滤功能可以针对具体绘图元素的具体某一项属性进行检索筛选（图11）。这个过滤功能在不同的 CAD 程序中有不同的叫法，有叫"过滤助理"（Filter Assistant）的，

实例：
　　如果两个图层中的成分交错重叠，虽然看起来依然错落有致，但是对各图层的编辑修改却存在一些困难。此时，可以在编辑其中一个图层的时候将另外一个图层锁定，避免错误操作带来的麻烦。

提示：
　　有些 CAD 程序（例如 Graphisoft ArchiCAD）会为建筑设计提供类似图层功能的楼层功能，专门用来绘制建筑设计中的楼层非常方便。

101

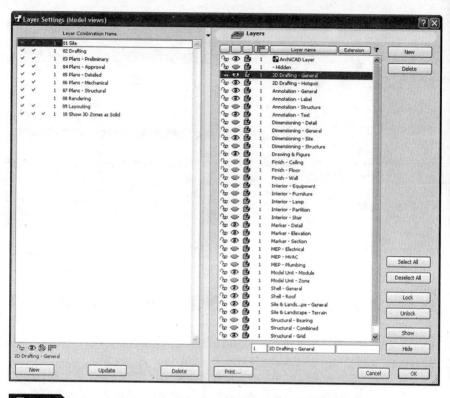

图10：
图层分级管理（Graphisoft ArchiCAD）

也有叫"查找及开启"（Search and Activate）的。无论怎样，这个功能在功能菜单里面很容易找得到。

实例：
在一个复杂的图层里有一套以特殊线型绘制的外墙，如果要对此外墙进行编辑，只需要通过过滤线型便可以找到这套外墙，而不必要——选择。

注解：
过滤功能可以针对各种CAD元素起作用，线型、颜色及成分特性（见"建筑元素"）都可以作为过滤条件。在绘制一类元素或一个组件时，可以采取某一种相同的过滤条件，以便日后查找编辑。

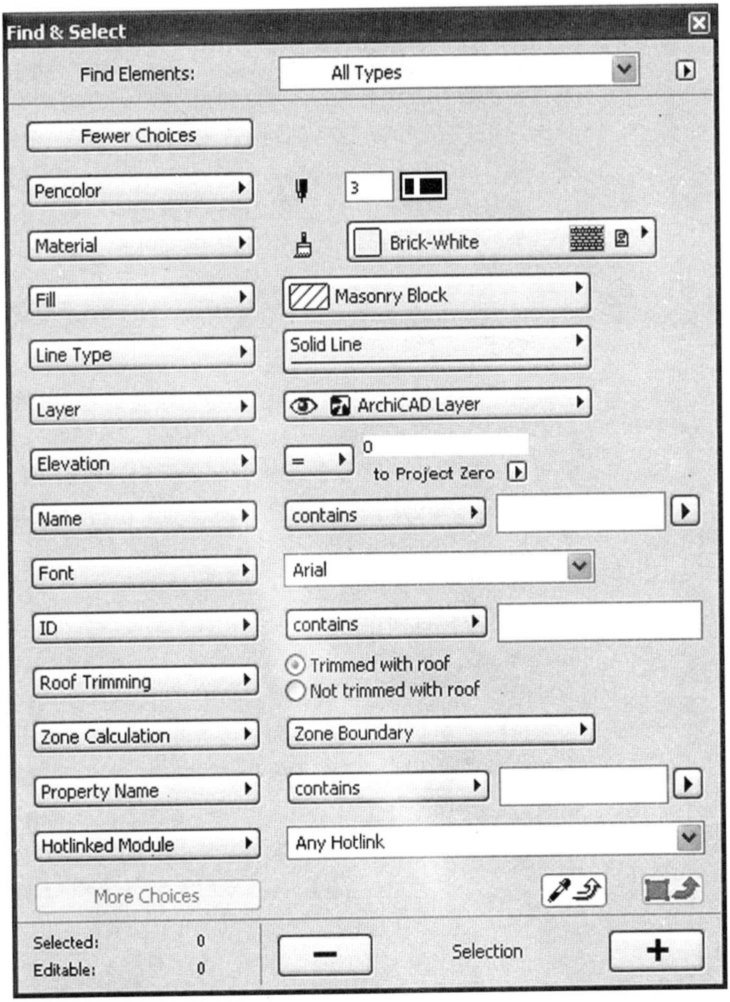

图 11:
各种各样的过滤条件 (Graphisoft ArchiCAD)

P22 **绘图功能**

本篇将基于一些简单的图例对一些基本绘图功能进行说明，其中所涉及的设计方法是浅显且通用的。诚然，在使用 CAD 时，我们可以通过很多不同的方法达到同一个目的，而对于一个使用者更重要的是，要针对自己的绘图找到一个最便捷、最有效的方法。

通过 CAD 的绘图命令可以很容易地绘制一些基本几何元素（点、线、方、圆），其他更多的绘图功能可以在用户界面的工具栏找到（见图 12 及"用户界面"）。

P23 **绘图元素**

点

与手绘图一样，在 CAD 绘图中，点是最基本的绘图元素。在几何学里，点是零维物体，不向空间的任何一维延伸。所有的几何图形都可以看作是由一定数量的点确定。例如，两点成线，三点成三角形，四点成四边形，八点成立方体等等。不同的 CAD 程序对点的显示有所不同（图 13）。

我们可以通过激活 POINT 命令来绘制点。

只需要用鼠标在绘图平面上一点，就可以定义一个点的位置并完成绘制，也可以在程序对话框中输入一个点的坐标来完成绘制——这两种绘制方法实际上在以下提到的所有绘制行为中都会用到。

线

绘制一条直线段只需要确定两个端点位置即可。我们用 LINE 命令在绘图平面上确定了一个端点位置之后，继续确定另外一个端点的位置，CAD 程序会在两个端点之间自动连接形成一条直线段（图 14）。

多义线

我们可以用这个多义线命令绘制一个连续的折线系（图 15）。

注解：

除了从工具栏和互动菜单中选择工具命令以外，还可以使用键盘快捷方式调用工具命令。最被我们熟知的是 Windows 系统中的 CTRL+C，选定对象并使用 CTRL+C 后，便可复制对象到剪贴板，供粘贴使用。每种 CAD 系统都会指定其各种工具命令的快捷方式，具体方法可以直接从菜单帮助中获得。

提示：

多边型是指包含一个闭合空间的多面体；多义线是指一系列的连续线段。

图 12：
绘图命令工具栏（Nemetschek Allplan）

样条曲线 　　　"样条曲线"这一术语来自造船业，用来描述与弯曲的船体锚接的具有同样曲率的肋板——"样条曲线"这一绘图功能同样可以达到描述类似效果的目的。在任意点之间内插弯曲圆滑的曲线是这一功能很容易做到的（图16）。

图 13：
确定点的方式

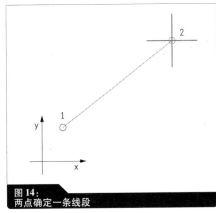

图 14：
两点确定一条线段

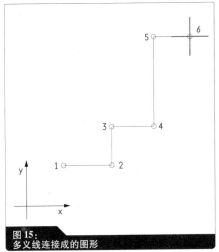

图 15：
多义线连接成的图形

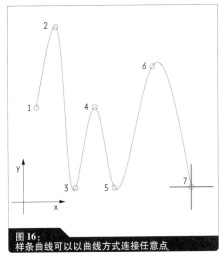

图 16：
样条曲线可以以曲线方式连接任意点

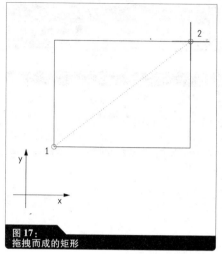

图17：
拖拽而成的矩形

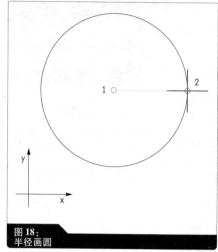

图18：
半径画圆

正方形和矩形	在CAD中创建一个矩形非常简单，只需要通过一个简单的功能就可以，而不需要逐个绘制矩形的四个边。我们只要确定矩形的对角线的位置，便可以方便地画出我们想要的矩形了（图17）。我们也可以通过鼠标的拖拽功能拉出一个矩形，还可以通过键盘输入相应的参数来获得我们想要的矩形。
圆	CAD中绘制圆的方法多种多样，最基础的方法是通过半径来定义圆。我们可以在命令对话框中输入CIRCLE命令，然后通过鼠标定位或者键盘输入参数来定义这个圆（图18）。在其他CAD程序中，可以通过定义直径或利用圆周切线定义周长上的3个点来画圆，将所有这些方法结合起来也可以达到画圆的目的（图19）。
圆弧	圆弧也是CAD中一个重要的绘图元素。应用CIRCLE命令就可以绘制一个完整的圆或者圆的一部分弧。当我们需要画一段圆弧时，首先确定好圆弧的中心和半径，然后定义圆弧的起始点和终止点即可（图20）。
椭圆	椭圆的绘制方法与画圆类似，惟一不同的是，画椭圆需要在确定圆心之后再行定义两个不同的半径（图21）。
椭圆弧	椭圆弧的绘制方法与画椭圆类似，追加定义椭圆弧的起始点和终止点即可（图22）。
剖面线、图案及填充	利用剖面线、图案及填充可以明确地显示绘图元素的相关属性，提高了绘图的可读性（图23）。剖面线主要用于显示设计部位的组成部分及材料情况；图案及填充可以帮助我们绘饰制图元素的表面，也可

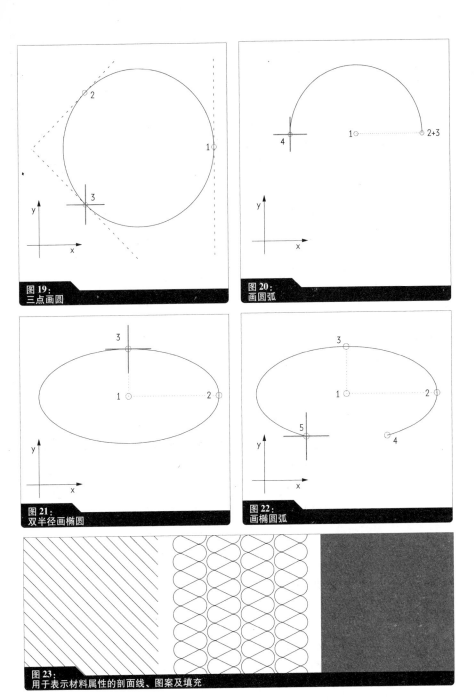

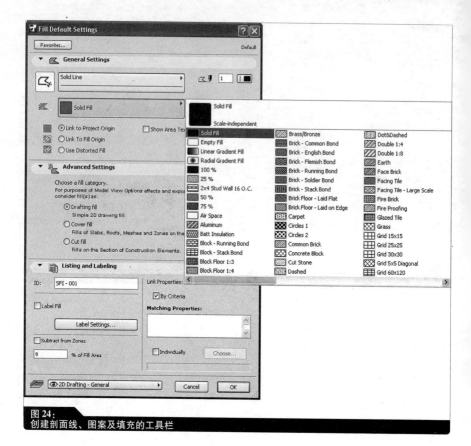

图 24:
创建剖面线、图案及填充的工具栏

以用在整个绘图的图示部分。不是所有的 CAD 程序都将这三种功能区分开来,许多 CAD 程序都是将这三个功能整合在一个工具栏里(图24)。

点击相应工具按钮,我们可以依次选取绘制区域的各个角点来确定绘制面,或直接定义一个矩形表面为绘制面(106 页)。

边界自动识别

许多 CAD 程序都提供了一种自动识别闭合图形边界的功能:如果一个图形的外边界完全闭合,我们就可以应用<u>自动识别边界</u>功能来识别图形边界,这个功能按钮很容易在工具栏中找到。

注解:
边界自动识别功能仅应用于完全闭合的范围。如果绘图时不够精确,轮廓线有可能存在微小的空隙,且在屏幕上以正常尺寸显示时看不出来。而试图发现这些错误也将耗费大量时间。因此,要想充分利用 CAD 的功能,精确绘图非常重要,何况许多绘图功能还是交叉应用的。

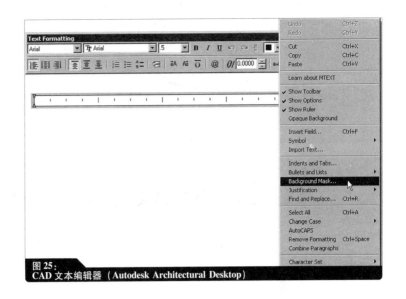

图 25：
CAD 文本编辑器（Autodesk Architectural Desktop）

在激活这个功能之后，CAD 程序便能够自动识别到我们所选图形的边界线。这个功能帮助我们节省了很多时间，在图形边界情况很复杂的情况下尤为明显。

文本元素

在许多 CAD 程序中，我们都可以应用文本编辑器在绘图中添加文本。与很多文字处理程序一样，文本编辑器提供很多用户自行设置的参数，包括字体及字号大小等等（图25）。激活文本编辑器之后，我们可以在文本编辑器中直接输入文本，也可以从其他的文字处理程序中拷贝文本到文本编辑器中（例如 Microsoft Word）。然而，有些 CAD 程序并没有提供专供输入文本的窗口，这种情况下，我们可以直接在绘图区域中输入文本，也可以直接选择绘图区域中的文本进行编辑。

标注线

我们在绘图时必须要对各个环节进行尺寸标注，在比例要求精确的技术制图时尤需如此。所有尺寸标注都可以通过标注链、高程点标注和个别标注等功能实现。

线性标注

线性标注功能可以对水平、竖直及对角元素进行尺寸标注。首先，不论次序地定义好要标注的控制点，在执行命令之后，CAD 程序会以一个固定的格式自动地画出标注线、箭头及尺寸文本（图26左侧部分）。所有标注线里面的组成元素的参数都可以通过对话框手动设定。控制点可以随意地增加和删除，程序会在增加或删除控制点之后重新自动计算并绘出所有标注线。

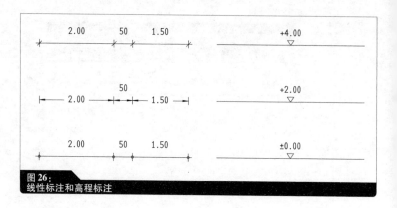

图 26：
线性标注和高程标注

高程标注　　高程标注同样需要定义控制点。高程标注在绘图中表现为一个等边三角形和一组数字。这类标注只是为了测量和显示高差而存在（图26右侧部分）。点击鼠标选择一个高程的基准点，然后再选择所要标注高差的控制点，CAD 程序会自动计算高差并将高差按照固定格式显示在控制点旁边。

P30　　　　**设计工具**

本篇对设计工具的介绍是在总结各种 CAD 程序的基础上的一些概念性总结，在不同的 CAD 程序中，大致相同的设计工具在不同的 CAD 程序中的叫法及配置方法可能会有所不同。在使用不同的 CAD 程序时，我们需要再次学习其特殊的功能及相应的配置使用方法。总而言之，选择一个适合自己使用的 CAD 系统不仅仅依赖于 CAD 产品的质量，还与自己的绘图需求有关。

栅格　　　栅格在设计过程中起到非常重要的作用，在设计较大对象时的辅助作用尤为显著。例如，栅格会使设计建筑物的承重结构及外观设计变得更加简单，仅需要借助对栅格排列的参照，便可以完成很多繁琐的绘图过程（图27）。

捕捉功能　　在绘制精度较高的绘图时，我们必须精确地定义绘图对象各个点

提示：
　　为了保证各高程点之间的正确联系，精确定义参考高程是必不可少的，以此为基础可得到不同的高程点。参考高程通常以建筑物底层的竣工地面标高（FFL）来表示，即 ±0.00m。关于标注线的详细附加说明可参见本套基础教材中的《工程制图》（征订号：18811，中国建筑工业出版社 2010 年出版）。

的坐标（见"坐标系"），这些坐标需要通过键盘输入到 CAD 程序中。显然，通过鼠标选点要比输入坐标值效率高些，这时我们便需要"捕捉功能"来帮忙了。在菜单栏或相应的工具栏里可以找到"捕捉功能"，这一功能可以保证我们在选取、定义点的位置的时候丝毫不差。"捕捉功能"配合前面提到的"栅格"在一起使用是很方便有效的，开启捕捉功能，当我们把鼠标指针移动到栅格点旁边时，CAD 程序会自动将鼠标指针定位到栅格点。我们执行 LINE 命令画线，线段的起始点及终止点都可以利用捕捉功能选取（图28）。这一功能很切实地提高了绘图精度。

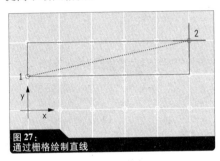

图 27：
通过栅格绘制直线

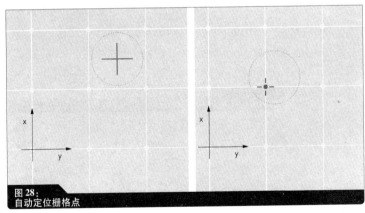

图 28：
自动定位栅格点

注解：
　　栅格即点阵，用户可以定义点之间的距离，点之间可以用线联系。例如，栅格可由均为 1m 间隔的 X、Y 轴组成。CAD 中，栅格也可以像用户坐标系（UCS）一样任意移动。

提示：
　　由于绘图元素与坐标系是紧密联系在一起的，所以精确性对于 CAD 绘图来说非常重要。即便是在开始时的一个小错误，也会随着绘图工作的进行渐渐变成大错误。

捕捉功能通常是跟其他的绘图命令一起使用的，在执行其他绘图命令之前必须要提前激活捕捉功能，并设置好捕捉半径。

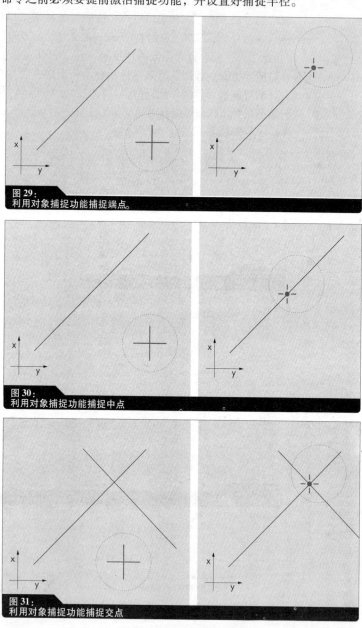

图 29：
利用对象捕捉功能捕捉端点。

图 30：
利用对象捕捉功能捕捉中点

图 31：
利用对象捕捉功能捕捉交点

对象捕捉　　　　捕捉点的形式多种多样，可以是线段的起点、终点或中点，也可以是多边形的端点，还可以是圆的圆心或两个绘图对象的交点等等（图29～图31）。显示这种点的方式跟前面提到的捕捉功能类似，鼠标指针移动到捕捉点附近时，会自动移动到捕捉点位置，并以图示形式告知。捕捉半径可以由用户自行定义，意为捕捉功能生效时鼠标指针与捕捉点的最大距离。当被捕捉点进入到代表捕捉范围的虚圆内时，捕捉功能自动识别被捕捉点并准确定位。代表捕捉范围的虚圆在屏幕上的半径通常有几毫米左右。

角度　　　　在进行精度较高的技术制图时，精确的角度定义是非常重要的。CAD程序在定义角度方面有几种不同的方法。通常情况下，我们可以使用键盘在对话框或角度选择器中交互输入我们所需要的角度。在很多CAD程序中，我们也可以在绘图时通过按住Shift键调用事先定义好的角增量来进行角度定义。这类辅助工具可以在CAD工具栏里找到（图32～图35）。

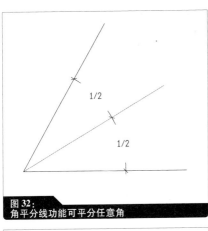
图32：
角平分线功能可平分任意角

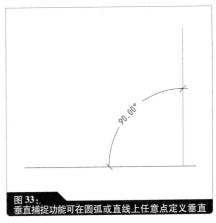
图33：
垂直捕捉功能可在圆弧或直线上任意点定义垂直

图34：
平行线功能可创建与选定直线平行的任意直线

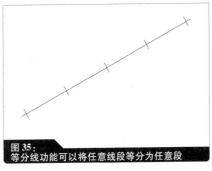
图35：
等分线功能可以将任意线段等分为任意段

修改功能

CAD 的一个很大的优势在于允许制图者随意编辑、修改绘制好的绘图。这意味着在整个设计过程中，我们可以随时对设计绘图进行优化，极大地节省了工作量。如果有必要的话，我们可以修改设计模版，使模版适合不同的需求，这样，我们每次开始绘图的时候便可以调用已经绘制好的绘图对象，稍作修改便能适应新的绘图需求而不必重复劳动（见"程式库"）。

关联性

CAD 各元素的修改是具备关联性的。比如，我们对一个正方形填充了剖面线，在修改正方形形状的时候，剖面线也会针对新形状进行适应性修改（图36）。这说明正方形与其剖面线在几何形状上已经构成了关联，一个改变，另外一个也会跟着改变。另外一个例子是我们前面提到的标注线，当被标注对象改变时，其标注线以及与其标注线关联的标注链都会自动改变，以适应新的绘图对象。这意味着我们不必重新对改动过的绘图对象进行标注，CAD 会自动根据关联性修改原有的标注链。

群组

在 CAD 中，绘图对象通常都是由很多绘图元素组成，这些绘图元素虽然都在一个绘图对象中，但是彼此之间并不一定都是关联在一起的。比如，一个绘图对象为一张床，这张床由好几条线绘制而成，当我们要修改这些线的时候，必须一个一个地修改。我们可以将这些绘图元素组合成一个群组，以便我们在选择绘图对象的时候不用一一对其中的绘图元素进行选择而直接选择群组便可。依靠 CAD 这一功

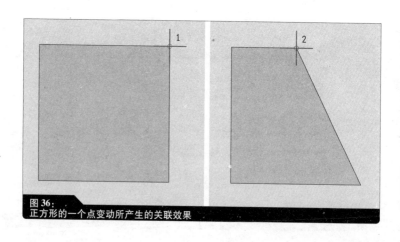

图36：
正方形的一个点变动所产生的关联效果

> **注解：**
> 拉伸一个绘制好的建筑图形非常容易，只要选择一个边上的几个关键点，移动这些关键点时，待拉伸边被相应地拉伸。

> **提示：**
> 当我们做其他操作的时候也能够用到窗选功能，窗选功能应该是我们最常用的选择功能之一。

能，我们可以将绘图对象中的各绘图元素创建为"群组"、"块"或"节"，所创建的这类单元在需要的时候也是可以被拆散还原的。这样一来，设计工作便简单多了。

关键点修改

在CAD中通过一个点或几个点修改绘图对象是非常灵活的一件事，我们用鼠标就可以修改一条线段的长度，也可以很容易地修改平面及三维绘图对象等（见"三维绘图"）。

在激活相应命令之后，我们可以使用捕捉功能捕捉到矩形的一个端点，然后移动该点到想要的位置，以达到我们修改绘图形状的目的（图36）。

擦除

我们在工具栏中可以找到擦除功能，该功能可以擦除绘图中的任何绘图元素，分别擦除或批次擦除均不在话下。在激活擦除功能之后，我们可以用鼠标在绘图空间中窗选需要擦除的绘图元素。根据设定的不同擦除方式也会有所不同，我们可以擦除选择窗选区域内的所有绘图元素，也可以删除凡是跟窗选区域相交的绘图元素（图37）。

裁剪

使用裁剪功能可以移除绘图元素的一部分或多部分。这一功能也叫"截断"或"删除点间线"，激活这一功能之后，我们可以裁掉一条直线在相交边界线以外的部分（图38）。

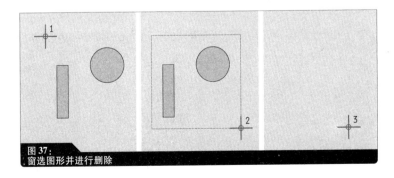

图37：
窗选图形并进行删除

延伸

复制

与裁剪相反，延伸功能（也叫"加长"或"连接"）可以将一条线段延伸到想要与之相交的那条边界线（图39）。

复制功能同样能够大幅提高绘图效率。一般而言，在重新绘制一个绘图对象之前，我们应该仔细考虑一下这个绘图对象是不是可以被复制然后进行编辑，如果可以，那么我们以复制加再编辑这一方法代替重新绘制绘图对象会非常地节省时间（见"程式库"）。

首先选择好需要复制的绘图对象，然后将其复制，再将副本拖拽至新的目标位置，原来的绘图对象位置保持不变——复制过程便是这样完成的（图40）。

镜像

可以使用镜像功能创建一个与原绘图对象以任何一个轴镜像对应的新绘图对象，这一过程中，原绘图对象可以保留，也可以不保留。在激活镜像功能并选择好原绘图对象后，需要依据与原绘图对象的距离和角度确定一个对称轴（也可以利用已经绘制好的直线或绘图对象的边界来作这个对称轴）。

与复制功能相配合，镜像功能可以很方便地创建左右对称的绘图对象，当然，我们必须先要画好对称结构中的一半，然后再选择相应的轴进行镜像操作（图41）。

移动

移动功能与复制功能非常相似，惟一不同的是，绘图对象是被移动到了新位置，而不是被复制的（图42）。

旋转

应用旋转功能可以将绘图对象以任意点为中心按任意角度进行旋转。如果我们只是希望改变绘图对象的方向也很简单，只需要把旋转点捕捉在绘图对象的中心点，然后设定方向角度便可。我们所说的方

注解：
可以对一条或多条线使用裁剪或延伸功能。因而，可以同时选取、裁剪或延伸任意数量的线条。

注解：
捕捉点还可以用来抓取图形对象。你可以十分精确地捕捉到图形对象并将它移动至另一位置。同时，捕捉点可用于定位。

实例：
当我们绘制楼层平面布置图时，对于对称结构便可以用到镜像功能，这样非常节省工作量。同样的方法可应用于快速有效地绘制建筑物立面图和构件图。

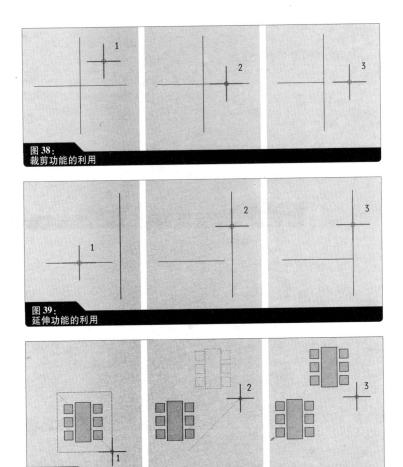

图38：
裁剪功能的利用

图39：
延伸功能的利用

图40：
复制一套餐桌

向角度一般是指用户坐标系 X 轴正向与原始坐标系正向所成的角度，在确定好旋转中心后，我们用鼠标就可以随意改变这个角度。当然，我们还可以用键盘输入旋转角度。比如，我们把转折点设定在绘图对象的端点，然后把转折角度设定为 90°，绘图对象便会直立起来（图43）。

缩放及拉伸　　所有绘图对象的大小、长度和形状都是基于精确的比例进行改变的。缩放和拉伸功能便能达到这类要求，我们可以应用这种功能将绘图对象沿着坐标系任何方向按照一定的比例进行缩放（图44）。

117

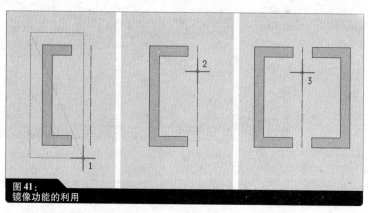

图 41：
镜像功能的利用

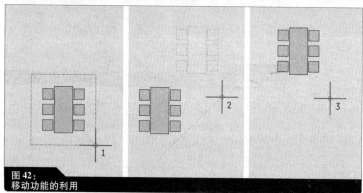

图 42：
移动功能的利用

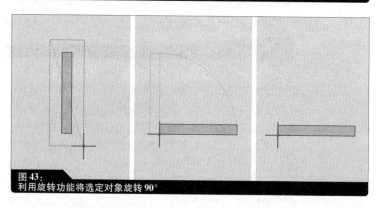

图 43：
利用旋转功能将选定对象旋转 90°

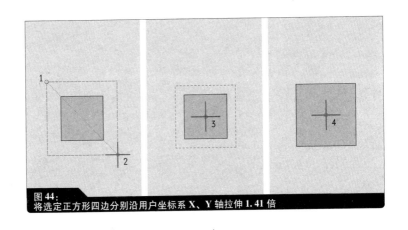

图44：
将选定正方形四边分别沿用户坐标系 X、Y 轴拉伸 1.41 倍

P40

三维绘图

前面章节介绍了二维绘图中的一些基本的绘图功能，这些绘图功能给 CAD 平面绘图提供了很大的帮助。CAD 还有一个很值得称道的长项，那便是三维绘图。

普通的绘图只能显示绘图对象的平面效果，而 CAD 的三维设计功能能够帮助设计师创建绘图对象的三维绘图。此时，为了能够将平面绘图扩展到三维空间，需要有另外一个参照系加入到原有坐标系中去，Z 轴便应运而生。我们最常应用的坐标系为笛卡儿坐标系。在此坐标系中，Z 轴通过原点并垂直于 XY 平面，本书第 99 页之图 5 中 P (3，5) 点可以在三维坐标系中重新定义（图45）。

P40

三维设计

在进行三维绘图时，原来的二维绘图平面可以看作是三维绘图空间的一个组成元素，而在三维绘图空间中的一切绘图操作行为都是基于原有二维绘图平面的。这样一来，三维绘图实际上可以看作是二维

注解：
三个坐标轴的方位遵循"右手原则"，即大拇指和食指分别代表 X 轴和 Y 轴，中指垂直于大拇指和食指伸出的方向便是 Z 轴方向（图46）。

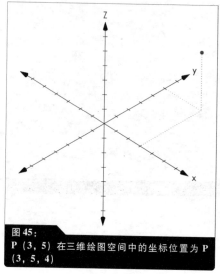

图45：
P（3，5）在三维绘图空间中的坐标位置为 P（3，5，4）

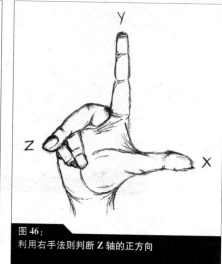

图46：
利用右手法则判断 Z 轴的正方向

投影

绘图的一个延伸。

二维绘图只能在平面上表现绘图对象（例如平面布置图或视图），而三维绘图能够是给我们展现各个不同立面，甚至立体视图（图47）。所有的展现方式都针对同一个三维对象，这样可以清晰地从多个视角了解三维绘图对象，并且使我们在编辑绘图时清楚地知道任何一个小改动在不同视角上产生的变化。大部分 CAD 程序会在计算机屏幕上提供多个视窗用于展现不同视角，我们可以从不同角度观察绘图对象。

模型

三维绘图对象根据设计方法的不同有着不同的定义，大致可以分为体积模型、平面模型和边缘模型几类。

体积模型

一个立体实体反映在 CAD 绘图中时，将会包含很多绘图信息（图48）。体积模型不仅可以展现三维实体，还可以提供诸如质量、重心等关键数据，并能够对实体的表面特性及其他具体数据进行定义（见"视觉效果"）。这样可以使在虚拟状态下展现三维实体的结构组成等关键属性成为可能，而且，展现实体的三维图像是可以进行修改的，只需要修改体积模型的几何参数便可做到。这样一来，多个体积模型之

注解：
在开始三维设计之前，我们必须将原来的二维思维扩展到三维思维，只有这样才能避免不该犯的错误。用纸板搭建简易抽象的三维模型进行实体比对是一个行之有效的方法（见"构图方法"和"视觉效果"）。

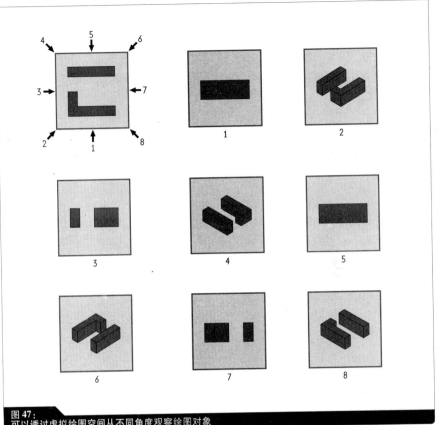

图47：
可以透过虚拟绘图空间从不同角度观察绘图对象

平面模型

边界模型

间的交叉操作（联合、删剪或交叉）也是可以实现的（见"建模"）。

顾名思义，平面模型仅针对于平面而言，而不涉及体积方面的定义。在CAD程序中，平面可以依据绘图对象表面的视觉特性进行定义，这些平面在虚拟绘图区内可以任意旋转或偏向于任何角度（图49）。

边界模型是指在三维绘图空间中用线段来描述绘图对象的一种模型，边界模型不包括表面或体积等信息。边界模型由于其高度抽象的特性，在CAD中仅为辅助模型而用（图50）。

> **提示：**
> 多个平面模型闭合起来形成的模型与体积模型看上去很相似，但是确切地说，更应该是一个空腔型的模型，描述为一个体积模型的外壳也不为过。

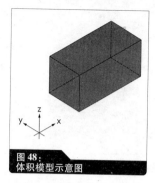
图48：
体积模型示意图

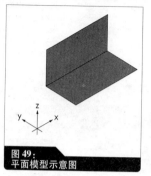
图49：
平面模型示意图

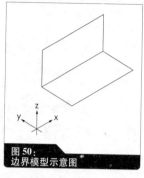
图50：
边界模型示意图

线框模型　　　在大多数 CAD 程序的三维设计模式中，所有的空间对象都是以线框模型的方式表现出来的，这样是为了能够将所有的点坐标清晰地表示出来。

　　区别于上面提到的边界模型，线框模型能够更加详细地以线段描述三维绘图对象的形状轮廓。在这一设计模式下，所有的平面模型和体积模型都以线框模型描述，当然，此时的线框模型中还是不包含平面和体积信息的。在虚拟绘图空间中，我们可以通过线框模型上的任意一个点来选定线框模型。由于线框模型即便是在二维绘图平面上表现时也能够表现出空间性，所以有时在复杂绘图中也是需要区别线框模型的正反面的（图51）。有些 CAD 程序并不提供线框模型功能，而通过直接对表面进行遮光或颜色处理而达到相同的目的。

边界隐藏　　　在需要时，三维绘图对象所有的边界都可以隐藏起来。边界隐藏之后，整个绘图的样式会好看很多，也更加容易让人理解，当然，这个时候的样式照比渲染之后还是差很多的，还有更重要的一点，此时的绘图是可以很容易就修改的（图52 及 "视觉效果"）。其他 CAD 程序称这种功能为 "隐藏"（HIDE）或 "隐藏线"（HIDDEN LINE）。

> 注解：
> 　　计算边界隐藏实际上是非常费时间的，为了提高效率，我们应该在开始绘制三维模型时就考虑到降低三维模型的细节深度。而当我们需要一个突出细节的三维模型时，我们把细节部分安排在其他图层，这样便可以分立地进行边界隐藏运算了（"透明的平面——分层原则" 及 "视觉效果"）。

> 提示：
> 　　有许多 CAD 程序在设计模式下都提供了预览窗口，设计师可以在预览窗口里看到隐藏边界、阴影处理甚至表面着色之后的三维模型，并且这个窗口里面的预览三维模型是可以旋转多角度观察的。

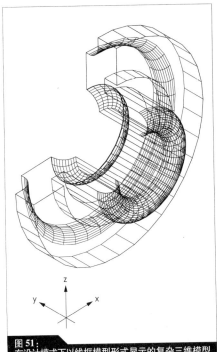

图 51：
在设计模式下以线框模型形式显示的复杂三维模型

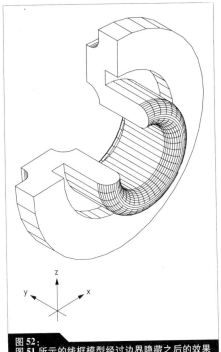

图 52：
图 51 所示的线框模型经过边界隐藏之后的效果

生成平面布置图、视图及截面

平面布置图和横断面

工程制图通常要求详细绘制出整个结构或结构的不同组成部分的平面布置图、视图及截面。在三维设计模式下，我们可以先建立一个完整的虚拟模型，必要时，这些模型可以作为生成二维图的基础模型。

在默认模式下，CAD 的绘图平面为一个顶视平面，也就是说，使用者正视用户坐标系（UCS）的 XY 平面从而看到虚拟模型。为了能够表现虚拟建筑的平面布置图，我们需要生成一个横断面，即以一个水平面在某一高度处横切建筑物而得到的截面（图 53）。

提示：
图 53～图 55 所示的房屋结构取自坐落于 Vaucresson 由柯布西耶设计的公寓（1922 年），尺寸和细部与原图稍有出入。

注解：
横断面平行于用户坐标系的 XY 轴平面，纵断面平行于用户坐标系的 Z 轴。有些 CAD 程序（例如 Graphisoft ArchiCAD）更提供了按照楼层提取横断面的功能，设计人员可在同一建筑的不同楼层平面图之间交替工作。（"透明的平面——图层"一章）。

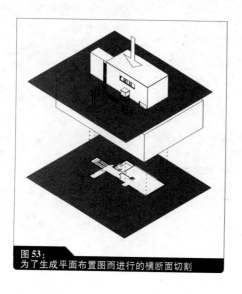

图53：
为了生成平面布置图而进行的横断面切割

视图和纵断面

大多数 CAD 程序都会使用同样的工具来生成视图和截面。技术制图中，断面也被称作视图，即切过建筑物或物体，能使我们看到内部结构。与生成平面布置图相反，切面为纵向或侧向时便生成了纵断面。我们可以通过确定切面位置来确定纵断面视图，然后再确定投影面。CAD 程序自动在我们确定好的投影面上绘制出二维纵断面图，我们可以直接对这一二维纵断面图进行编辑修改等操作（图54、图55）。

P48

构图方法

三维绘图的构图方法与二维绘图构图方法类似，所不同的是，三维绘图比二维绘图多了一个 Z 轴（见"三维设计"）。

空间中的点、线、面

与我们在二维绘图平面上绘制点、线、面时差不多，我们也可以在三维绘图空间的用户坐标系中绘制点、线、面。在三维绘图空间绘图时，绘图元素并不一定总是平行于坐标轴，于是我们会经常对绘

> **提示：**
> 有些 CAD 程序（例如 Graphisoft ArchiCAD）可以在绘图区域提供三维视图的多个投影图区域，所有绘图变动都会在各个投影图区域内产生变化。相应地，投影图可在另外的图层上仅由二维线段生成。

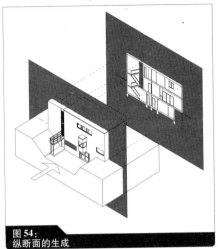

图 54：
纵断面的生成

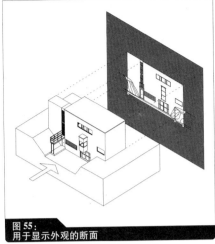

图 55：
用于显示外观的断面

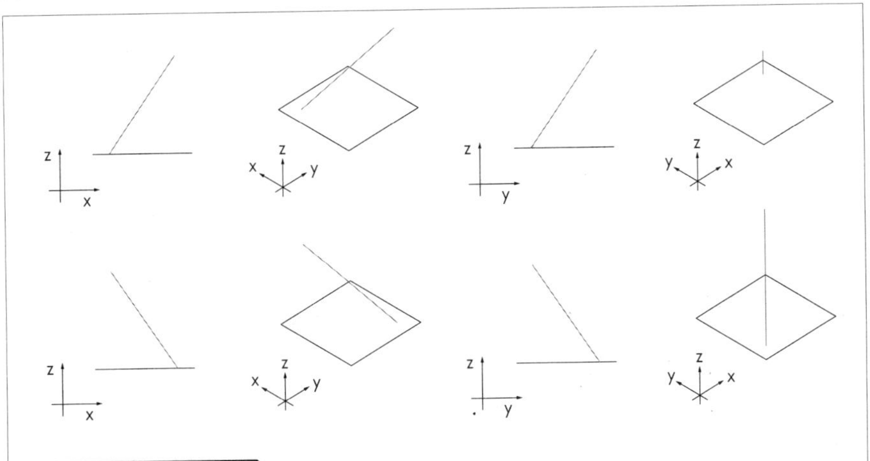

图 56：
从三维空间的不同角度观察一个平面和一条直线

元素的长度等特性产生错觉（图56）。

注解：
在三维绘图空间，可以生成多边型平面。不仅可以在竖直及水平方向生成平面，还可以生成倾斜的平面。当生成倾斜平面时，只需要在二维绘图平面上先生成一个二维平面，然后在三维绘图空间中以一定角度旋转就可以了。

长方体 　　我们利用体积模型的一些绘图功能可以很方便地生成简单的单体。例如，想要生成一个长方体，我们只需要确定其底面积，然后再定义好它的高度便可以生成了（图57）。地面通常基于平行于用户坐标系的 XY 轴平面的平面来设定。

圆柱体 　　圆柱体可以是正圆柱体，也可以是椭圆柱体。定义好底面的类型，再给出圆柱体的高度，圆柱体便可生成（图58）。

圆锥体 　　我们定义圆锥体的时候需要确定一个或为正圆或为椭圆的底面，然后利用 Z 轴确定一个顶点位置，便可生成一个圆锥体（图59）。

球面体 　　在 CAD 中创建球面体，需要在球面体的赤道切面上确定好球面体的中心点以及赤道面半径或直径。球面体的所有纬度平面都是平行于用户坐标系的 XY 轴平面的，中轴平行于 Z 轴（图60）。

挤压出来的体和面 　　上述一些三维绘图功能使我们很容易地创建简单的几何立体模型，但是我们需要注意的是，并不是所有的 CAD 程序都提供了类似的

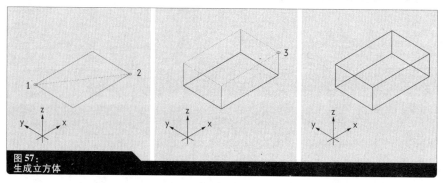

图57：生成立方体

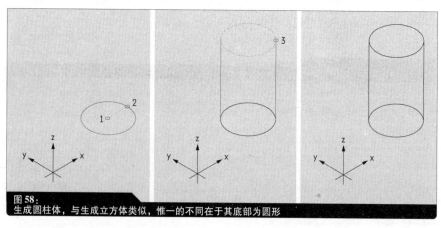

图58：生成圆柱体，与生成立方体类似，惟一的不同在于其底部为圆形

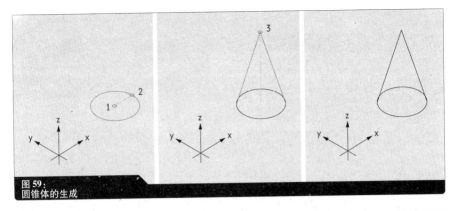

图59：
圆锥体的生成

功能，即便有了这些功能，创建立体模型也会受到截面几何形状的制约而难于实施。有个好办法可以解决这些问题，我们把这个办法叫做"挤压成型"功能。把平面图上的几何形状的闭合区域想像成一个模具的出口，成型物质从这个出口被挤压出来迅速成型，便形成了我们想要的立体模型——这与做生日蛋糕上的奶油裱花有些类似。有了这个功能，我们可以将二维平面上包括直线、多义线及样条曲线等各种图形挤压成型为体和面。

这一方法很适合用于有倒圆角、斜面以及其他特殊几何形状的绘图对象的绘制。需要注意的是，二维平面图形应该是闭合且不与其他图形交叠的。如果二维平面图形不闭合，我们还不如直接定义一个立体平面，这样更简单些。

"挤压成型"功能在不同的CAD程序中有不同的叫法，例如"拉高"（RAISE）、"变换"（TRANSLATION）或"投射"（TRAJECTION）。"挤压成型"过程是沿着图形的边界线或穿过闭合区域而进行的，二维平面图形的基本形状为边界线所构成的边界轮廓，沿着边界线伸出的方位线叫做"伸出路径"（PATH）。有些CAD程序要求同时定义基本形状和伸出路径（例如Nemetschek Allplan），而有些CAD程序（例如Autodesk Architectural Desktop）只需要确定基本形状和输入伸出方位线端点的坐标即可完成操作（图61）。

注解：
挤压成体的方法用于生成三维地形图具有一定优势。每条等高线都对应一个图层，图层再根据等高线的高程确定标高，按照这个思路便可以很容易地生成三维地形图了。关于手动建模的详细信息可参见本套基础教材中的《模型制作》（征订号：18843，中国建筑工业出版社2010年出版）

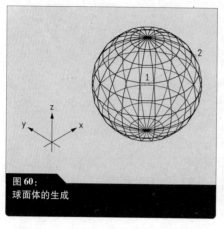

图60：
球面体的生成

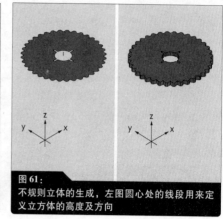

图61：
不规则立体的生成，左图圆心处的线段用来定义立方体的高度及方向

旋转而成的体

我们利用"旋转"功能可以将一个二维平面图形围绕任意空间轴旋转。与"挤压成型"功能类似，"旋转"功能同样可以简化截面形状复杂的立体模型的创建过程。反映截面形状的二维平面图形同样需要是闭合且不与其他图形交叠的。如图62所示，位于XY轴平面的剖面围绕一个平行于Y轴的直线进行旋转，得到了图示中的三维模型。这一过程与制陶有些类似，陶坯放在一个定轴旋转的底座上，随着底座的旋转及施加在陶坯上的外力变化便可得到想要的陶坯模型。

P52

建模

前面讲了几种生成三维模型的方法，其实，三维模型也是可以进行再造型的，也就是说，三维模型的基本形状是可以被修改的。我们可以通过改动三维模型各个端点在用户坐标系中的位置来改变三维模型的形状。例如，只需要几个步骤，我们便可以将一个长方体变成一个楔形体（图63）。或者，经过一系列复杂的调整，我们可以改变一个已经绘制好的建筑物模型的总高度。

三维模型的组合搭建

造型其实并不是简单地调整几个点那么简单，有时会用到很多命令及方法。例如，三维模型之间是可以互相组合以及拆分的，进行三维模型组合及拆分的绘图命令在不同的CAD程序中叫法不同。有叫"组合"（COMBINATION）的，有叫CSG（"几何体构建"的缩写）的，也有叫"布尔操作"（BOOLEAN OPERATIONS）的。实际上，这些操作基本上都是以"增加"、"削减"及"交叉"为基础来进

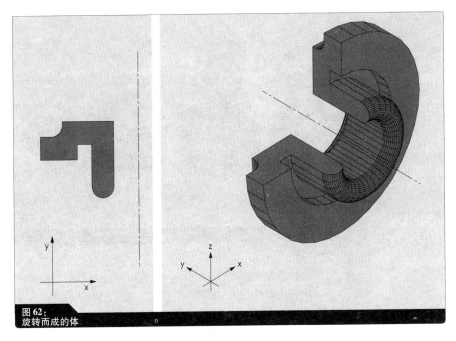

图 62：
旋转而成的体

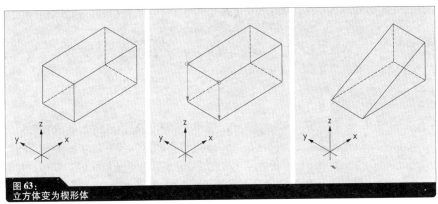

图 63：
立方体变为楔形体

行的。

增加 　　"增加"可以将两个或多个三维模型合并为一个三维模型，此时，多个三维模型相互增加合并的顺序并不重要。在这些三维模型组合为一之后，便成为一个整体，即可以看作是一个群组。这一操作在将多个建筑组成部分组合成为一个建筑整体时起到很重要的作用（图64b）。

129

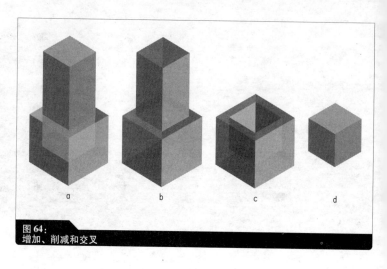

图64：
增加、削减和交叉

削减　　　　　　我们可以利用"削减"来达到将一个三维模型从其所在三维模型群组里消除的目的。在这一操作中，削减对象的选择顺序是很关键的，只有第一个被选择的三维模型在削减操作中保持原封不动，即与被削减三维模型相交的部分会维持原样。被削减的三维模型被以原有的形状移除，从现实角度理解，更像是将三维模型群组拆解开来（图64c）。这一操作使生成切槽形状的图形变得非常简单。

交叉　　　　　　我们可以将"交叉"理解为削减三维模型交叉部分以外的部分而形成新的三维模型的过程（图64d）。

P54　　　　**建筑元素**

　　　　　　一般情况下，CAD中的图形元素都是根据几何形状或各点的坐标进行定义，所有这些几何信息使我们能够计算与之相关的几何尺寸。另外，我们还能将这些几何信息利用在其他特定的需要中。我们可以利用CAD程序将一些例如建筑图例等特定的图形和几何形状定义为通用的绘图用建筑元素。例如，许多复杂的建筑中有很多诸如墙体、窗户和楼梯等组成部分是可以被事先定义好的，这样一来，这些组成部分不需要每次都重新绘制，将绘图元素调用出来进行重新定义便可。所谓的重新定义，即指重新设定图形的线型、剖面线以及表面特征等相关属性（见"视觉效果"）。当然，我们还可以将这些事先定义好的绘图元素分类存储，方便查找调用（见"TAI"）。如此，我们可以很有效率地进行绘图，并能够准确地统计绘图信息。

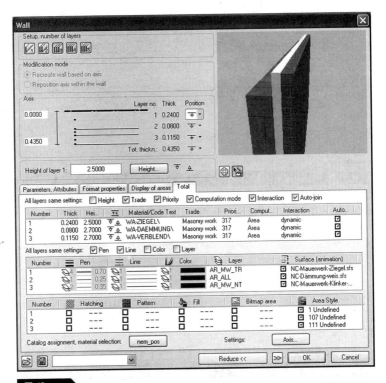

图 65：
墙体工具的对话框（Nemetschek Allplan）

绘制建筑
元素

> 墙体

与绘制二维图形类似，建筑元素通常都是以顶视图的模式在二维平面上绘制。与二维绘图本质的不同在于，我们需要在绘图时加入更多的参数，例如结构层、高度以及材料等参数。

当我们使用"墙体"工具绘制墙体时，需要在对话框中定义各种与墙体有关的参数。（见图65）

我们可以使用默认的设置来定义墙高及结构，从而生成单层或多层的墙体，每层墙体都可以用不同的线型来表示，有必要的话，还可

提示：
将很多有所联系的建筑物构件信息合并在一起进行统一处理，对日后的工作会非常有利。例如，构件搭接处或多层墙体间的材料连接可自动生成，无需耗费太多人工。

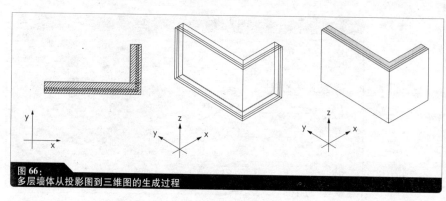

图66:
多层墙体从投影图到三维图的生成过程

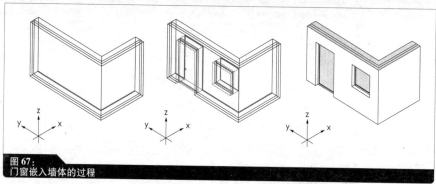

图67:
门窗嵌入墙体的过程

以定义视觉属性以及进行数量计算。定义完所有这些参数后，便可以按照我们的意愿来生成复杂的图形了（图66）。这样一来，我们便能够通过使用一个简单的绘图功能便可以完成步骤繁琐的二维甚至三维绘图了。

孔、窗和门

在绘制完墙体部分后，我们可以在墙体上任意设置门和窗。与简单地定义开孔长宽不同，我们可以直接将一套成型的门或窗整体嵌入到墙体中（图67）。

提示：
　　墙体开孔和嵌入墙体的建筑组成部分（门、窗等）都是联系在一起的，此类建筑组成部分均可以自动适应墙体开孔，完成嵌入过程。

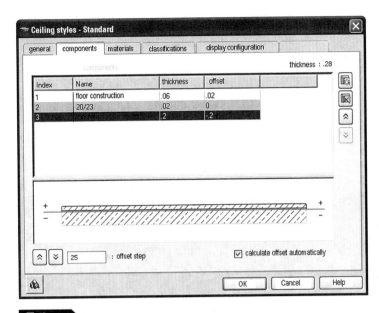

图68:
楼板工具的对话框（Autodesk Architectural Desktop）

楼板　　　　　与墙体的生成方法类似，生成楼板也需要事先定义好剖面线及表面特性，然后确定形状及面板的厚度和位置。需要的话，还需要确定与总量计算及视觉效果处理有关的所有属性参数（图68）。

　　　　　　　只要按照上述步骤定义好楼板样式，我们便可以在绘图区域中方便快捷地绘制楼板了。开口在楼板上的主图楼梯等组成部分可以以多义线的形式在楼板平面上绘制。

楼梯　　　　　手绘楼梯是件费时费力的工作，利用CAD绘制楼梯便简单得多了。楼梯的生成主要依赖于楼梯平面投射图的长宽以及层间高度，与之相关的所有参数都可以在生成过程中甚至生成之后进行改动。我们只要选定好楼梯的样式，接下来所有有关楼梯生成的细节设定都可以根据程序对话框的提示进行（图69）。

　　　　　　　在完成参数设定之后，CAD会自动计算并生成楼梯，我们可以

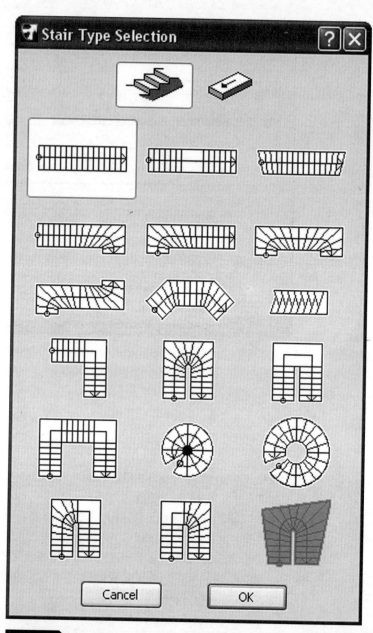

图69：
楼梯样式库（Graphisoft ArchiCAD）

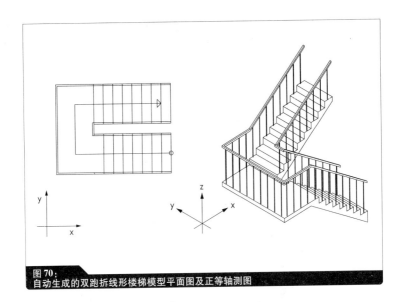

图70：
自动生成的双跑折线形楼梯模型平面图及正等轴测图

屋顶

在三维视图中核查图形的细节，必要时还可以进行修改（图70）。

屋顶的几何参数大致由房屋底部断面形状、屋顶轮廓及倾斜度、屋脊、边缘高度及屋檐等因素决定。很多CAD系统提供了绘制简易屋顶的工具，使用多义线确定房屋底部断面，并在对话框中输入所有必须的参数，便可直接生成所要的屋顶（图71、图72）。

提示：

有些CAD程序（如Graphisoft ArchiCAD）提供了完整的楼梯样式库，使用者可以从样式库里直接调用楼梯样式；其他CAD程序（如Nemetschek Allplan）则以定义好的二维平面图为基础，通过用户对集合参数及结构参数的自定义来生成楼梯模型。

注解：

许多CAD程序都提供了自动进行屋顶结构添加的功能，也就是说，设计师在确定好屋顶样式及相关几何参数后，CAD会自动添加椽梁等结构部件。

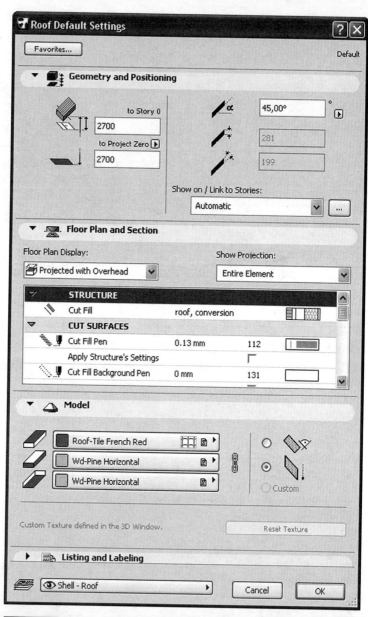

图71：
屋顶工具的对话框参量与线型可一起定义楼层高度、屋顶倾斜度等。（Graphisoft Allplan）

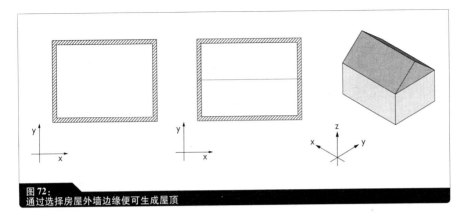

图72:
通过选择房屋外墙边缘便可生成屋顶

视觉效果

"视觉效果"在建筑学领域专指以绘画形式表现建筑物细部的一种形式。在计算机的帮助下，我们可以进行更加高级的视觉效果模拟，从而使更多的非专业人员更直观地了解设计细节，这样显然比研读一项一项的技术数据来得简单。

计算机视觉效果模拟有很多用处。设计师既可以利用这一设计工具来比对设计对象的结构元素比例是否恰当，也可以衡量整个设计对象与周围元素之间的空间关系。另外，在进行建筑工程方案规划时，视觉效果功能可以用于评价材料效果和照明效果。在设计工作初期，利用简单抽象的虚拟模型可以提前预知很多诸如材料适应性和结构比例等与设计有关的问题。由此，我们可以持续不断地通过完善虚拟模型来为设计工作提供决策帮助。一个个渐被完善的虚拟模型搭建在一起，构成一个完整的虚拟建筑，这对整个建筑设计的帮助是巨大的（图73）。

计算机视觉效果模拟同样可以帮助我们衡量不同设计的利害得失。比如，一个建造场所的外部环境条件不变，将不同的设计放置其

> 视觉效果作为设计工具

提示：
为了达到良好的视觉效果，我们必须要用到专用的渲染软件。很多CAD程序中都集成了渲染软件，但是仅仅这些集成的渲染功能很多时候是不能达到我们的要求的。

图73：
视觉效果作为设计工具，可在设计进程的不同阶段审查其体积、光线和材料

视觉效果用
于设计展示

中，便会直观地感觉到各个设计与外界环境是否协调（图74）。

与设计图和等比例模型一样，建筑视觉效果图同样可以用于建筑设计展示。针对具体要求以及不同观者，视觉效果图可以形象化些，也可以抽象些。一个好的展示图并不一定是极端模拟现实的，简单抽象的非现实手法同样可以展现建筑物的各种信息（图75）。

细节丰富的视觉效果渲染需要计算机进行大量的计算，很费时。

注解：
　　许多渲染软件允许使用数码图像作为视觉效果图的背景，这样会为视觉效果图增色不少。比如，用建筑物周围景观的数码图像作为建筑物三维视觉效果图的背景，会给人很逼真的感觉。

图74：
同一场景下的不同建筑物视觉效果图

如果可能，在进行细节视觉效果渲染之前应该生成一个虚拟模型，对虚拟模型进行大致的视觉效果渲染以确定合理性，最后再进行细节丰富的视觉效果渲染，这是一个省时省力的做法（见"三维绘图"及本篇"渲染参数"）。例如，在展示城市面貌的时候，某一栋建筑物的诸如突起、凹陷等等细部在视觉效果图里面就显得不那么重要，即便在图中绘出了也不会被看到（图76）。

在绘图视点距离建筑物绘图实体较近的时候，建筑物的细部表现才显得重要。对建筑物绘图实体周围的环境进行装饰，会使效果图更加生动，也会给观看者提供实际参照物，对被展示的建筑物有更全面

139

图75：
抽象型的视觉效果图

图76：
城市规划的视觉效果图

图77：
视觉效果图中的建筑物细节和周边环境细节的重要性

的把握（图77）。

室内空间效果展示也是建筑视觉效果展示的一个重要应用领域。室内空间效果展示偏向于细部展示，当制作一个虚拟现实的室内空间效果图时，我们需要注意各个结构细节和表面细节（图78）。

表面

表面光学效果处理是计算机视觉效果处理的重要组成部分，包括对颜色、纹理和光源效应等的处理。

颜色

我们之所以能够观察到物体的表面，是因为色谱效应的存在。光线到达物体表面时，会被吸收或反射。如果色谱中所有的颜色都被反射，那么我们看到的将会是白色；如果整个色谱颜色都被吸收，那么我们会看到黑色。由此看来，我们能看到什么颜色是取决于物体表面对颜色光的吸收及反射的。例如，物体表面为蓝色时，会反射光线中的蓝色而吸收掉其他的颜色谱。

模拟表面效果的一个重要方法就是有选择性地进行表面颜色设置。为了达到更逼真的模拟效果，需要对物体表面的光反射性进行非常细微的调整（图79）。

注解：
　　一个高质量的视觉效果图所表现出的虚拟现实效果是非常出色的，但是正因为这种出色，反而让人觉得不那么真实了。在进行视觉效果处理时，我们应该有意地根据现实情况对画面进行一些仿真处理，让效果图更加贴近现实而不是贴近"完美"。

图78:
内部空间视觉效果图演变

　　物体表面背光一面要比向光一面黑一些,另外,不管物体表面是什么颜色,强光线在其强反射面上的反射效果会偏白色。通常情况下,与定义镜面反射和透明度等参数类似,渲染软件会提供定义物体表面颜色和反射颜色的功能,需要的话,我们可以直接在程序对话框中输入百分比来定义。

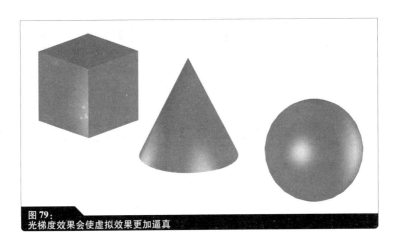

图79：
光梯度效果会使虚拟效果更加逼真

我们可以使用绘图用的电子笔来定义绘图对象表面的颜色属性和反射透明系数。在 CAD 系统中，绘图笔（或绘图笔颜色）是与绘图对象的种种视觉效果信息相关联的（见"虚拟绘图板"）。

纹理

纹理功能可用于表现虚拟对象表面的材料属性，我们通常使用位图贴图来达到这一目的。例如，一个砖墙的特写照片就是贴图用位图的一种（见"渲染参数"）。在 MAPPING 功能的帮助下，我们可以将位图贴在绘图对象表面并适应表面尺寸，经过渲染之后，绘图对象表面看起来就会像真的墙壁表面一样。将物体表面的局部照片引入到虚拟绘图中去，经过加工处理，是可以很好地模拟物体表面的实际视觉效果的（图80）。

P69

光影处理

良好设置的光源以及有选择性地加入阴影效果，对成功模拟物体的现实视觉效果是很重要的。渲染软件允许我们在虚拟绘图空间内部和外部设置和调整多个不同的虚拟光源。

实例：

对于一个透明的玻璃，光线是直接穿过玻璃而没有留下什么痕迹的，但是玻璃对光的反射及折射依然会对其周边的元素产生影响。因此，我们在进行视觉效果渲染时，我们应该多用渲染软件试验几遍，与脑海中的印象比对一下，这样能做出更好的作品来。

注解：

通常情况下，渲染软件里面都存储了很多标准的渲染用纹理，我们可以在手头没有其他选择的时候直接利用软件自带的纹理进行渲染。当然，我们在互联网上还可以搜索到更多的纹理。

图80：
供视觉效果渲染用的纹理

虚拟光源　　虚拟光源可以提供不同的光强和颜色，另外，还可以使正常情况下的可见光成为不可见，直到它指向某一平面——效果类似于透过雾气看到的火炬闪光。

聚光光源、平行光源和点光源等虚拟光源所提供的光效是不同的（图81～图84）。

环境光　　除了绘图空间内虚拟光源以外，远离绘图空间的光源也可能产生效果，我们称这种光效为环境光。环境光是一种发散光，充斥在绘图空间之内，有固定的照度和颜色。环境光越强，其他虚拟光源产生的效果便越弱。我们可以利用环境光来调整绘图空间的光效，避免光饱和或光浑浊效果的出现。

照明强度　　照明强度通过调整绘图空间内各个虚拟光源的色温来进行控制。色彩越明亮，虚拟光源的色温便越高，光亮度也就越高。纯白色的色温是最高的，所以光亮度也是最高。如果我们发现一个虚拟光源的光亮度不够，那么我们可以添加其他的虚拟光源进行补充。浅色光包含更多的光谱颜色，所以浅色光具备更好的环境颜色适应能力。

阴影　　与真实光源一样，虚拟光源产生的光照射在虚拟绘图对象上也会产生相应的阴影效果，这样看起来会更加逼真。根据计算方法的不同，可以生成不同效果的阴影（图85～图87）。

提示：
　　严格地说，太阳光是点光源，但是太阳光在经过了地球大气层的折射之后，变成了平行光。

注解：
　　有些CAD程序（如Graphisoft ArchiCAD和Nemetschek Allplan）具有智能阳光模拟功能。我们只要在对话框中输入现实地点的经纬度、年月日及具体时间，程序便自动模拟出与所有约束条件相符合的阳光、阴影效果。这是一个非常实用的功能。

P71　　　**透视图和虚拟相机**

　　选择正确的透视效果对于计算机视觉效果模拟是非常关键的（图72～图78）。视觉效果图必须考虑到所有与构图相关的因素，方能凸显细节且又具有吸引力。效果图内的各组成成分的显示也应该以现实世界的透视效果为准，这样，观察者在看到效果图时才会觉得逼真。在视觉效果处理过程中，我们可以将虚拟模型的各个组成元素与虚拟光源结合起来进行调整，以达到理想的透视效果。所有这些，组成了渲染场景。使用场景功能可以节省很多时间，我们可以把一些场景存储

图81：
光源在视觉效果图中的作用很重要

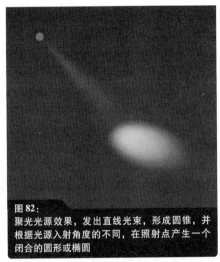

图82：
聚光光源效果，发出直线光束，形成圆锥，并根据光源入射角度的不同，在照射点产生一个闭合的圆形或椭圆

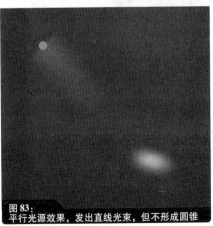

图83：
平行光源效果，发出直线光束，但不形成圆锥

图84：
点光源效果，非直线光束，向四周发散，类似于球状光照明

图 85：
线条硬朗的影子效果不需要太多的计算便可生成

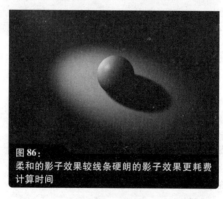

图 86：
柔和的影子效果较线条硬朗的影子效果更耗费计算时间

图 87：
模糊的影子效果最耗费计算时间

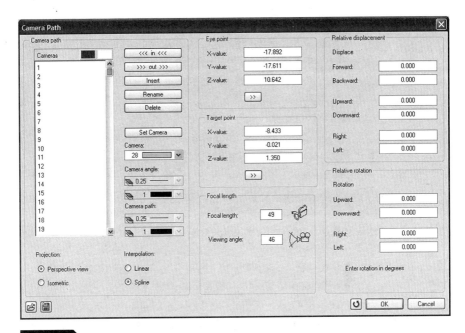

图 88：
虚拟照相机的参数设定

为模板，这样在每次视觉效果计算的时候就不用对同一场景进行重复计算了。通常情况下，我们会用虚拟照相机来捕捉渲染场景，虚拟照相机的诸如视点、对焦点、焦距以及镜头视角等参数都可以由绘图者设定（图88）。这里所说的"视点"专指在虚拟绘图空间中的观察视点。我们是可以通过三维坐标系来确定对焦点位置的（见"坐标系"）。

照相机视角和焦距在几何角度上是相互依赖的，两者决定了虚拟照相机的远景视野。即便在很狭窄的空间里，我们也可以通过减小焦距来获得更大的视野。然而，为了避免语言效应的出现，是不能将焦距设置得太小的。可能的话，视点和对焦点应该在同一高度上，这样会使各个垂直边缘看起来是平行的。

提示：
在使用渲染软件时，我们可以先将一个场景保存下来，便于日后调用。场景其实就是指虚拟绘图对象和虚拟光源的安排情况，可以通过虚拟照相机来获取场景（见"透视图"和"虚拟相机"篇）。

P75

渲染参数

视觉效果图的质量除了取决于绘图对象、表面处理和虚拟光源设置等方面以外,还要照顾到效果图本身的一些影响因素。协调设置与效果图本身相关的各种参数,同样是决定效果图质量好坏的关键。很多渲染软件都有预置好的参数,按照这些预置参数进行渲染是方法之一。下面对这些参数作一下大致的解释(图89)。

"渲染"(Rendering)这个词来源于其他领域,在计算机视觉效果处理领域中专指将矢量图转换为比特位图这一过程。

图形分辨率

一个位图的图形分辨率是以在一个单位区域内存在的像素数作为衡量依据的。正常情况下,图形分辨率以在 X 坐标上的每英寸点数(dpi)为单位。每英寸像素数越多,图形分辨率便越高。相反,每英寸像素数很少的位图的分辨率会很低,看起来颗粒感明显,较难进行细节分辨(图90)。

在像素总数一定的情况下,图形分辨率决定了位图尺寸的大小。比如,一个1000像素×1000像素区域,当分辨率为300dpi时的尺寸为8.47cm×8.47cm,而当分辨率减小到150dpi时,区域尺寸便会扩大为16.94cm×16.94cm。这样一来,原来那些像素被放置在比原来大4倍的区域里,每个像素都变大了4倍——这也就是为什么矢量图可以无限放大而不影响质量,位图却不行的原因。无限放大位图的后果只会让位图颗粒感越来越强。

效果图渲染计算的时间根据分辨率的不同而不同,简而言之,渲染计算中的像素数越多,计算时间便越长。在设计过程中所进行的效

提示:

有很多渲染引擎可供我们选择,各种渲染引擎都有各自的计算方法,目前主要有 Raytracing、Phong Shading、Gouroud Shading、OpenGL 和 Z-Buffer。为了区别各软件的渲染效果以便选择合适的渲染软件,最好还是多用这些软件进行实际渲染比对。

提示:

二维或三维矢量图中的各种绘图元素都是基于坐标系进行定义绘制的。与位图(见下文)不同,矢量图可以任意地放大缩小,而绘图质量却丝毫不受到影响。另外,与其他图像格式相比,矢量图所需要的存储空间也是相对较小的。

提示:

位图(也称像素图)是由一个个的像素排列组成的。像素是位图的最小元素,每个像素都包含一定的色彩信息,很多像素像马赛克一样排列在一起,便形成了位图图像。

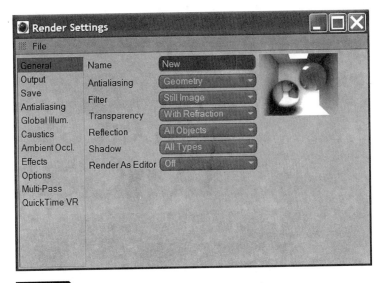

图89:
渲染设置对话框（Cinema 4D）

图90:
不同分辨率的渲染效果

果图渲染并不需要像最终效果图那样有太高的分辨率，这样会节省很多计算时间，提高工作效率。

反失真　　　　矢量图转换为位图的过程中，我们可以使用"反失真"功能来避

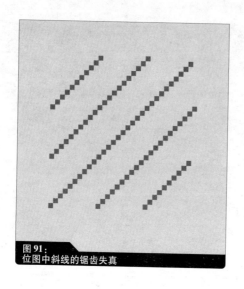

图91：
位图中斜线的锯齿失真

免在斜线上出现的锯齿失真（也叫阶梯失真）。水平线和竖直线在位图中是不存在失真现象的，而对于斜线来说，由于位图中的斜线是由很多正方形的像素偏移衔接而成，所以阶梯形失真却必然要出现的（图91）。

类似的失真情况在其他的类圆形状甚至在文字中都会出现，而且，图形分辨率越低失真越明显。反失真功能通过调整一些"阶梯"的色值来弱化阶梯形状，使它们看起来不那么明显。反失真的使用在某种程度上会加大计算量，效果图中的类圆形状和斜线越多，需要的计算量就越大。所以，最大化的反失真只在效果图的最后生成阶段才使用，这在计算量优化上显然是很有效率的。

除了大量的几何图形信息以外，虚拟模型渲染还有很多细节设置需要调整，这些调整对计算量大小的影响是非常至关重要的。出于对计算量及工作效率的考虑，我们在开始制作视觉效果图之前，必须先考虑好对效果图质量的要求到底有多高，然后按照质量要求量体裁衣，最优化地完成视觉效果图制作工作（附表2）。

注解：
　　可以通过提高分辨率的方法来减少锯齿失真，当然，高分辨率对应的是更多的像素，并需要更大的存储空间。

注解：
　　开始时，我们可以只对一个细部进行渲染，对虚拟场景、光影效果以及材料效果进行检查，确认之后再进行全局渲染。渲染一个复杂些的场景通常需要几个小时的时间。

应用数据集

程式库

CAD 程序通常都会提供一个整合了各种标志、文档模版和预置建筑组件的程式库，例如建筑类 CAD 程序就提供了各式各样的标准图供绘图者直接引用或修改后使用（图92）。

我们也可以将我们自己绘制的图中的某一个单元或整个图保存为模版，以便在另外一个图中使用，在引用这些模版的时候，我们同样可以对模版进行再编辑，以适应新的绘图要求。这样可以减少很多工作量，从而提高工作效率。

标志

程式库中有很多室内装饰元素（家具、洁具、橱具等）和环境设计元素（植物、汽车和人物等）。还有很多用于描述结构工程、机械工程及建筑学专业等的各类型元素的标志可供选择。

建筑组件

标准部件的应用可以很大程度地减少工作量，特别在绘制施工图时，我们可以很容易地从程式库中调取钢构件（梁、管等）连接件（螺钉、螺母等），稍作编辑修改后便可使用在绘图中。即便是诸如楼梯、门窗等复杂些的构件也都可以存储为标准件，供日后使用（见"建筑元素"）。

位图

除了矢量元素以外，位图也可以加入到程式库中。所谓的位图可以是数码相片、扫描图片，甚至渲染图也可以导入到程式库中供日后导出使用。

文档模版

整个绘图文档也可以导入到程式库中，也就是说，将一个绘图按照一定的比例尺存储为一个文档，并将此文档导入到程式库中，需要时，在配置好比例尺后可以在其他绘图中直接导入该文档。这样一来，有些诸如打印设置等繁琐的设置我们只需要在被调用文档中设置一次便可（见"打印和制图"）。

注解：
我们可以利用扫描仪将各种图形导入到 CAD 中使用；也可以将 CAD 中的绘图打印出来，以手绘图对 CAD 绘图进行修改，再用扫描仪将修改过的绘图导入到 CAD 中。这是手绘图与计算机辅助绘图良好结合的一个好方法。

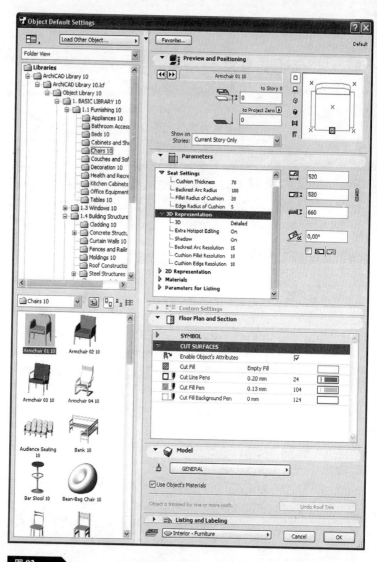

图 92
程式库（Graphisoft ArchiCAD）

CAD 对外接口

　　CAD 接口是与外部程序进行数据交换的桥梁。虽然所有的 CAD 系统都是使用矢量图作为绘图格式的，但是大多数 CAD 程序都有各自独立的文件格式，这些文件格式通常是互相不兼容的，这个相互之间的数据利用造成了障碍。DXF 文件格式是一个被大多数 CAD 系统认可的文件格式，大部分的 CAD 程序都可以读取和编辑这种文件格式，甚至有些图像软件也可以读取这类文件。

DXF 是<u>绘图数据交换格式</u>（Drawing Exchange Format）的缩写，是由 Autodesk 公司开发出来专门用于绘图数据交换的。

当一个 DXF 文件被导入到 CAD 程序中时，会有一些诸如单位、比例尺和二维三维转换等选项供我们选择设置，以适应新矢量图的要求（图93）。

很多建筑 CAD 程序不仅支持 DXF 文件格式，还同时提供了其他对外接口来避免数据交换可能带来的麻烦。这在使用渲染软件时是非常重要的，渲染效果图经常会应用到其他辅助软件，这对制作一个完美的效果图几乎是必须的。例如：Nemetschek Allplan 使用一种特殊的数据格式导出 3D 模型以供 Cinema 4D 进行渲染；Autodesk Architectural Desktop 为 3D STUDIO MAX/VIZ 提供了一个兼容的数据接口；Graphisoft ArchiCAD 为 Artlantis 专门开发了一个数据接口，等等。

TAI

TAI 是招投标（Tendering）、决标（Awarding）和计价（Invoicing）的英文首字母缩写，代表了建筑建造领域的一个重要环节。在建筑师完成设计和建造方案之后，投标或报价工作便会紧锣密鼓地展开。所有与设计建造有关的信息都会以一个统一的格式派发给投标者，各投标者根据自己的实际情况给出报价。接收到各投标者的报价，并对比各个投标者给出的价格和质量标准之后，便可以选择合适的投标者进行工程委托了。工程交付后，所有发生的费用要做成清晰的帐目。

提示：
数据导入导出实际上经常出现问题。某一 CAD 系统中的重要细节在作为原始信息输出以后经常被丢失或不能显示，这主要是由于各个系统在单位、字体、标注信息等方面存在差别而造成的。

提示：
有些特殊的程序可以读取和写入很多种文件格式，这时，利用 DXF 文件进行数据的导入导出便不是那么必须了。但是，想要作到这一点，还需要作很多的准备和实践方才可以。

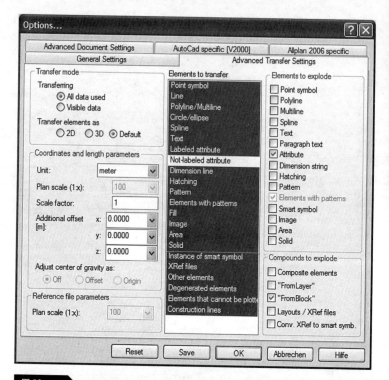

图93：
导入导出 DXF 文件（Nemetschek Allplan）

TAI 程序可以让以上工作变得非常简单。当设计师使用 CAD 创建虚拟建筑模型的时候，便可以将所有建筑组成成分根据原料和工时进行分类整理，供工程量计算使用（见"建筑元素"）。所有这些信息都可以直接导入到 TAI 程序当中并可继续进行编辑，从而省却了很多人工投入。

CAD 系统、TAI 软件、施工工地之间的联系

理想状态下，设计建造数据由 CAD 图中导出到报价程序，在这一过程中，工程量将自动与建筑组成成分联系起来，我们可以分别在 CAD 程序和 TAI 程序中核对相关信息。这使我们能够很有效率地编写合同条款，并能行之有效地监视工程进度。所有从 CAD 程序中导出的信息（属性、数量等）都可以用于计算并组织生成对应的目录。

打印和制图

在设计模式下完成制图之后,我们可以将图打印或进行制图(出图)。小尺寸的图可以在 A4 或 A3 的打印机上打印,大尺寸的图则要用到绘图仪。

设计绘制好的图通常要合理地进行页面安排之后才能正式打印出图,我们可以在一个操作视窗中完成页面安排操作。

打印比例

在打印之前,我们需要根据图纸要求来确定打印比例和打印纸张的大小,同时还需要确定与打印比例相关的线宽和字体大小。我们需要注意的是,图的参考比例要与打印比例相匹配,在保证图纸精确清晰的前提下,避免图纸尺寸过大或过小的情况出现(见"虚拟绘图板")。

以建筑平面图为例,当需要在图纸上既要显示外观部分又要显示相应细节时,我们必须在图纸上为外观显示和细部显示确定不同的比例尺,以对应不同的显示需求。

纸张规格

为适应展示要求,平面图可使用各种不同的长宽比。在打印图时,为了适应不同纸张的规格(例如 ISO/DIN A 等),我们可能会改变原图的长宽比以适应纸张。DIN A3 和 A4 是非常常用的纸张规格,很多打印及复印机器都支持这类规格的纸张,我们打印图纸时,多选择这类纸张会很方便。

纸张上的图纸区域由内图框界定,外图框可以认为是裁图基准线。图纸标头供描述图纸信息而用,例如制图者、出图时间、比例尺等等(图94)。

图框和图纸标头都有模版可供使用,我们只需要在使用之前确定好诸如比例尺之类的设置便可(见"程式库")。

虚拟打印

我们可以使用虚拟打印功能将绘图输出为可用于打印出图的文件。

提示:
绘图仪是指专门进行大尺寸打印的仪器。绘图仪可以打印 DIN A1 或 A0 的纸张,或可以打印宽度一定但是长度不限的纸卷轴。卷轴的宽度为 61.5cm(A1)或 91.5cm(A0)。我们可以直接在 CAD 里定义卷轴的宽度。

注解:
我们利用绘图仪打印图时,可以设置打印精度,通常情况下有三种打印精度可供选择。低精度打印速度很快,但是分辨率差,不过会比较节省墨粉。

提示:
每种 CAD 程序都提供了不同的平面设计模式,少部分 CAD 程序(Nemetschek Allplan)专门为平面设计和打印方案开发了很多的应用模块。

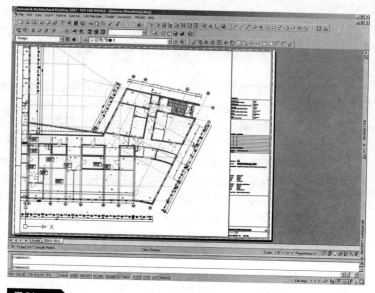

图 94：
图框（Autodesk AutoCAD）

PDF

位图输出

在这一过程中，绘图的矢量信息被虚拟打印程序转化为可供其他打印出图设备识别的文件，这个文件可以在没有 CAD 系统的支持下在其他打印机或绘图仪上打印出图。

　　PDF 文件格式是由 Adobe Systems 开发出来的文件格式，是 Portable Document Format 的首字母缩写。PDF 文件格式可以包含矢量图和位图，许多 CAD 系统都支持创建这一文件格式，并以此文件格式作为对外数据交换的途径之一。另外，通过安装虚拟打印机（例如 Adobe Acrobat Distiller）来生成 PDF 文件也是一种行之有效的方法。

　　有一小部分 CAD 系统可以直接输出位图文件。我们知道，位图文件有许多文件格式，我们可以根据需要来决定导出文件的格式，比如 JPEG 或 TIFF 等等。JPEG 是一种压缩位图文件格式，文件较小便于存储；TIFF 比 JPEG 要大，但是包含的信息量也大，在诸如图形设计等对位图精度要求较高的场合会用得到。

注解：
　　这种方法的应用比较普遍，比如，一个学生并没有自己的打印机，他只需要使用虚拟打印功能将图打印为文件，然后去寻找合适的打印机打印即可，重要的是，打印机所连接的电脑上并不需要有专门的 CAD 程序便可以打印。

P85 **系统需求**

P85 **硬件设备**

作为一个电脑用户,虽然不需要知道电脑硬件的具体工作原理,但是也应该要掌握一些硬件的基本知识,这对更好地利用电脑很有帮助。

中央处理器 中央处理器(CPU)是电脑的心脏和大脑,所有的数据运算都要通过它来完成。

主板 主板是众多电脑硬件设备协同工作的基础平台,上面可以连接CPU、存储设备、显卡、声卡及网卡等硬件设备。有一些电脑设备也可以集成安装在主板上,凡是集成安装在主板上的设备都可以称为"板载设备"。

内存 内存的英文首字母缩写为 RAM,即 Random Access Memory,是电脑的工作存储设备。

显卡 显卡是电脑的视频输出设备,电脑大部分的视觉计算工作都是由显卡来完成。对于 CAD 系统应用来说,我们最好不要应用板载显卡来进行视频输出,因为利用板载显卡来进行视频输出会拖累到内存从而降低硬件系统工作效率,间接影响到其他运算过程。

硬盘 硬盘是信息存储的主要介质,信息在被数据化以后写入到硬盘的磁性介质上,达到存储的目的。即便硬盘可以存储大量的信息,我们还是应该尽量把重要的文件刻录在 CD 或 DVD 上进行永久存储,保证万无一失。

显示器 显示器的大小通常用屏幕对角线的长度界定,单位为英寸。为了达到更好的显示效果,我们应该尽量选择大些的显示器。另外,在显卡支持的前提下,我们还可以架设两部显示器,一部用来显示工作区,另外一部显示功能界面。

特殊输入设备 除了鼠标、键盘以外,还有一些其他的输入设备可供选择,例如用于游戏的游戏手柄就可以派上一些用场。

空间鼠标 空间鼠标与游戏手柄类似,可以用来控制虚拟视觉效果的变幻。空间鼠标上还有按钮,可以用来执行某些绘图命令。

手写板 手写板也可以叫做数字化仪,是一种可以用电子笔绘图的输入设备。手写板的输入精度要比一般鼠标高,而且用起来与我们日常使用

触摸屏

的钢笔书写很相像。在设置好适当的比例之后,手写板的输入区域便可以看作是 CAD 的绘图区域了。

触摸屏是一种新型的输入方式,与手写板有些类似。当使用触摸屏进行输入操作时,整个屏幕既是显示设备又是输入设备,我们可以直接在屏幕上进行指点操作,当然,把它放平下来操作会更舒服一些。

P86

软件

硬件设备就绪之后,剩下的便是软件的选择。在老旧的硬件设备上运行新型的软件系统是件很令人头痛的事,选择一个适合硬件设备的操作系统是最需要事先考虑的。比较流行的操作系统有 Microsoft 的 Windows 系统和 Apple 的 MacOS,免费的 LINUX 也越来越被人们关注。操作系统与 CAD 系统有一定的兼容排斥性,在选择的时候应该注意软件生产者的特别要求。

市场上有很多种 CAD 软件可供选择,由于功能各不相同,在价格上会有很大差异。免费软件、共享软件及商业付费软件林林总总,层出不穷(附表3)。

有很多价格昂贵的 CAD 软件都会提供演示版或学生版。学生版的价格会比正式版便宜很多,而演示版更提供了一段时间的免费试用期。但是,这些版本还是有很多功能限制的。演示版大部分都只提供少数几个功能,仅供用户体验、评价而用。

软件选择

选择一套 CAD 系统需要考虑几方面因素:功能要求、使用要求、成本以及硬件需求。另外,数据交换也是不得不考虑的问题,在协作要求比较高的领域,尽可能采用支持同一文件格式的 CAD 系统是必须的。

除了学院教材以外,还有很多关于 CAD 的参考书籍可供选择。当然,针对个人情况参加培训课程也是可取的。

提示:

不同的软件厂商会对其软件产品的系统需求进行详细的说明,要注意的是,软件厂商所说的系统需求都是软件运行的最低需求。想要运行通畅的话,还是尽量提高系统配置的好。

注解:

免费软件和共享软件在互联网络上可以下载到,但是功能上都会多多少少地有些限制。

学习 CAD 应该是融会贯通的一个过程，掌握基本原理的基础上触类旁通，工作中会更加灵活。虽然各种 CAD 程序的操作有所不同，但这些程序都有着相同的基础构成，只要掌握了最基本的思路，便可以很容易地掌握所有 CAD 程序。

与电脑交互进行设计

利用电脑进行 CAD 绘图极大提高了设计工作及数据管理工作的效率，良好的虚拟现实性能及设计参考性能已经帮助人们确立了全新的设计理念。但我们不能否认，由于设计师与设计图纸之间存在着电脑这一实际屏障，有很多事情还是需要克服。从使用铅笔绘图转变到使用鼠标等输入设备进行绘图，这一过程确实需要一些努力和时间。对于初学者来说，有很多操作方面的问题需要面对，复杂的操作流程很有可能给初学者带来很多障碍。

从另一方面看，手绘图是直觉转换为作品的一个过程，在这一过程里往往有悄然天成的灵机一动，而在电脑上绘图却难以找到这种感觉。所以，我们还是提倡在使用 CAD 绘图的过程中尽量合理地加入手绘图，这对设计工作是有好处的。

CAD 目前已经成为设计师必不可少的工具，将来的发展更是不可限量。在建筑领域，虚拟模型的应用为所有的工程参与者提供了与设计工作直接联结的可能，这使得整个建设工作从设计环节到施工环节均能合理有效地协调联动起来。在工程进行时，设计能够随着工程的进展随时进行调整，根据实际情况进行设计优化，互动高效地完善每一个流程。所有与结构计算和几何运算相关的复杂流程对于 CAD 来说都不在话下，对于使用者来说，把它看成是一个简单易用的绘图工具就好了。在今后的建筑设计领域，必会越来越多地看到 CAD 的影子。

P89　　附录

P89　　软件要点表

建筑设计图层结构示例　　附表 1

环境		建筑物
		外延
		树木
		附属物
设计		设计栅格
		场地
		视图
		截面
承载结构		结构栅格
		基础
		外墙
		内墙
		柱
		梁
		楼梯
		楼板
		屋顶结构
建筑配套设施		电网
		照明结构
		电气安装
		暖通
内部装饰		浴室
		家具

续表

环境	建筑物
	物品
	地板
绘图信息	标注
	标签
	房间
	房间标记
	剖面线
	图案
	填充
	标记
	注释
	修正点
平面图	框架
	插图
	扫描图
	渲染图

计算机视觉效果处理要点列表 附表2

能力	硬件性能
	软件条件
	时间因素
视觉效果图用途	设计工具
	阶段性展示
	最终展示
视觉效果图等级	近距离观察
	建筑示意
	远景展示（例如空中摄影）
渲染场景	远景选择及虚拟照相机设置

续表

能力	硬件性能	
	图像格式	
	渲染对象	
	光景	
	背景及周边环境	
渲染参数	渲染类型（例如 Raytrace Phong Gouraud，OpenGL，Z-Buffer 等）	
	分辨率	
	反失真	
表面处理	颜色	
	纹理	
	反射效果	
	透明效果	

CAD 软件列表　　　　　　　　　　　　　　　　　　附表 3

程序名称	主页
Allplan	www.nemetschek.com
ArchiCAD	www.graphisoft.com
ArCon	www.arcon-software.com
AutoCAD/Architectural Desktop/Inventor/Revit Building	www.autodesk.com
BricsCad IntelliCAD	www.bricscad.com
CAD	www.malz-kassner.com
CATIA	www.catia.com
ideCAD	www.idecad.com
MicroStation	www.bentley.com
Reico CADDER	www.reico.de
RIB ARRIBA ® CA3D	www.rib-software.com
SketchUp Pro	www.sketchup.com
Spirit	www.softtech.com
Solid Edge	www.ugs.com
TurboCAD	www.imsi.com

渲染软件列表　　　　　　　　　　　　附表4

程序名称	主页
3ds Max/VIZ	www.autodesk.com
Artlantis	www.graphisoft.com
Cinema 4D	www.maxon.net
Maya	www.alias.com
mental ray	www.mentalimages.com

P92　　图片来源

图1，图4a，图12，图65，图88，图93：	Nemetschek AG, Munich-Allplan 2006
图2，图3，图4b，图10，图11，图24，图69，图71，图92：	Graphisoft-ArchiCAD 10
图4c，图9，图25，图68，图94：	Autodesk GmbH Deutschland-Architectural Desktop 2007
图53－图55：	Bert Bielefeld, Isabella Skiba Sonja Orzikowski
图73，图74，图76，图77，图78：	ch-quadrat architekten
图75：	Frank Münstermann
图81：	HKplus architekten
图82－87：	HKplus architekten/the author
图89：	Maxon Computer GmbH-Cinema 4D R10
其他插图：	The author

尊敬的读者：

感谢您选购我社图书！建工版图书按图书销售分类在卖场上架，共设22个一级分类及43个二级分类，根据图书销售分类选购建筑类图书会节省您的大量时间。现将建工版图书销售分类及与我社联系方式介绍给您，欢迎随时与我们联系。

★建工版图书销售分类表（见下表）。

★欢迎登陆中国建筑工业出版社网站www.cabp.com.cn，本网站为您提供建工版图书信息查询，网上留言、购书服务，并邀请您加入网上读者俱乐部。

★中国建筑工业出版社总编室　　电　话：010—58934845　　传　真：010—68321361

★中国建筑工业出版社发行部　　电　话：010—58933865　　传　真：010—68325420

E-mail：hbw@cabp.com.cn

建工版图书销售分类表

一级分类名称（代码）	二级分类名称（代码）	一级分类名称（代码）	二级分类名称（代码）
建筑学（A）	建筑历史与理论（A10）	园林景观（G）	园林史与园林景观理论（G10）
	建筑设计（A20）		园林景观规划与设计（G20）
	建筑技术（A30）		环境艺术设计（G30）
	建筑表现·建筑制图（A40）		园林景观施工（G40）
	建筑艺术（A50）		园林植物与应用（G50）
建筑设备·建筑材料（F）	暖通空调（F10）	城乡建设·市政工程·环境工程（B）	城镇与乡（村）建设（B10）
	建筑给水排水（F20）		道路桥梁工程（B20）
	建筑电气与建筑智能化技术（F30）		市政给水排水工程（B30）
	建筑节能·建筑防火（F40）		市政供热、供燃气工程（B40）
	建筑材料（F50）		环境工程（B50）
城市规划·城市设计（P）	城市史与城市规划理论（P10）	建筑结构与岩土工程（S）	建筑结构（S10）
	城市规划与城市设计（P20）		岩土工程（S20）
室内设计·装饰装修（D）	室内设计与表现（D10）	建筑施工·设备安装技术（C）	施工技术（C10）
	家具与装饰（D20）		设备安装技术（C20）
	装修材料与施工（D30）		工程质量与安全（C30）
建筑工程经济与管理（M）	施工管理（M10）	房地产开发管理（E）	房地产开发与经营（E10）
	工程管理（M20）		物业管理（E20）
	工程监理（M30）	辞典·连续出版物（Z）	辞典（Z10）
	工程经济与造价（M40）		连续出版物（Z20）
艺术·设计（K）	艺术（K10）	旅游·其他（Q）	旅游（Q10）
	工业设计（K20）		其他（Q20）
	平面设计（K30）	土木建筑计算机应用系列（J）	
执业资格考试用书（R）		法律法规与标准规范单行本（T）	
高校教材（V）		法律法规与标准规范汇编/大全（U）	
高职高专教材（X）		培训教材（Y）	
中职中专教材（W）		电子出版物（H）	

注：建工版图书销售分类已标注于图书封底。